An Introduction to Piaget

P. G. Richmond

Senior Lecturer in Education,
St Osyth's College of Education, Clacton

Basic Books, Inc., Publishers

NEW YORK

Library of Congress Catalog Card Number: 73-116854
SBN 465-09514-3
Printed in the United States of America
10 9 8 7 6 5

Contents

An introduction to Piaget

Introduction

In recent years Piaget's researches and theories have been recognized by educators as having a considerable bearing upon their activities. Several books have been written, all of which have made Piaget's work more widely known. A number of these which are particularly valuable are listed further on, under the heading, 'Other Sources of Reading in Piaget's Psychology.' It seems to the present author that the breadth and depth of Piaget's studies is such that there is room for many derivative works, each of which can make a useful contribution to the wider understanding of the significance of the psychology. However, it will be of help to the reader if, at this stage, some general points are made about how this book is orientated, and what basic principles underlie its format and content.

Professor Jean Piaget is most widely known as a child psychologist; nevertheless he is also a zoologist, a mathematician, and a philosopher; but perhaps, above all, he is a genetic epistemologist. This abiding interest in the processes by which bodies of knowledge grow, historically and as internally organized systems, has considerable bearing upon the whole structure and content of his child psychology. Piaget's study of children's thinking is part of wider issues which arise in genetic epistemology, and this fact explains many of the unique characteristics which are found in Piaget's study of intellectual development. It explains, for example, the stress upon the categories of reason, like qualitative classes and quantitative relations, objects and space, causality and time. It explains the stress upon mental organization and logical operations, and upon the growth of such structures from birth to adolescence. It also explains the biological components of the theory, since, in the end, the working of the intellect depends upon biological organization. Thus Piaget's research into the development of children's thinking is concerned with biological organization and with developing structures and their logical operations as they relate to the 'principal "categories" which intelligence uses to adapt to the external world—space and time, causality and substance, classification and number, etc.'.

Clearly, it would be most misleading to suggest that this short book can convey more than an outline of these facets of Piaget's psychology, for it would be impossible to condense a work of such magnitude into the confines of these pages and yet remain completely faithful to Piaget's meaning. However, it seems to the author that it should be possible to convey the flavour of Piaget's child psychology by attempting to abstract the main ingredients outlined above, and then combine them proportionately. To do this, there would be a need to outline the biological aspects of the theory, the structural and operational aspects, and the environmental factors which are at work in the developmental process. It is the aim in this book to try to present in a simple way this complex balance of ingredients without too much distortion of the whole psychology. Also underlying this simplification, there is the general principle that the empirical and theoretical content should have enough coherence to make it applicable to the learning situation so that those concerned with the education of children can think about how the psychology might be usefully applied. In order to explicate these aims the book has the following format.

Section one describes the process of intellectual development with an emphasis upon the individual grappling with his environment through various stages. The nature of this activity, its content, and the form of the mental actions which develop, are discussed. The broader theoretical principles are kept to a minimum in this section. Section two, on the other hand, looks at the broader principles underlying the developmental process. The biological roots of the intellect are outlined, the structures and their logical properties are discussed, together with some of the factors which influence their development. It is a moot point whether Sections one and two can be meaningful when taken separately, since each rests on the other. However, taking both sections together, there has been an attempt to build up the theory and terminology in a gradual way, so that the reader is eased gently into the abstractions of the psychology. Section three rests upon the preceding sections and is, of course, largely speculative. The first half attempts to transplant the theory into the learning situation (i.e. Section two applied); while the second half relates the developmental sequence to educational content and practice (i.e. Section one applied). It is inevitable that Section three will result in a narrow view of both the psychology and the learning situation. As far as the educational side is concerned, the singularity of the approach must display the learning situation in a strange perspective. On the other hand it may also throw into sharper relief some otherwise less clearly defined features. The speculations which make up Section three will have served a useful purpose if they raise questions in the reader's mind. To look at old matters in a new light is always worthwhile. Sections one and two will have served a useful purpose, also, if the reader is encouraged to turn directly to Piaget's writing.

At the end of the book, under the heading of 'Problems and Difficulties', some mention has been made of areas of discordance between Piaget's developmental system and other researches. These comments are in no way a critical assessment of his work, but they have been included to temper the conclusions outlined in Section three.

I am greatly indebted to Dr K. Lovell, B.Sc., M.A., Ph.D., Reader at the Institute of Education of Leeds University, for his incisive comments which have led to a number of improvements in the manuscript. I am indebted also to Mr F. C. Lamb, B.Sc., M.Ed., and Miss D. Harris, M.A., Principal Lecturers in Education at St Osyth's College of Education, for their comments and suggestions during the preparation of this book. I am especially indebted to my wife, Jean, for the many hours she has spent in typing and retyping the manuscript. I am grateful to many others with whom I have discussed questions which arise from Piaget's developmental system, and not least the teachers and student teachers who grapple with the complexities of the psychology.

St Osyth's College P. G. R.

Key to references made to Piaget's work

Reference has been made to the following works by Piaget, all of which are available in English translation. The Key below indicates the short form of the titles of these books which will be found in the text.

FULL TITLE AND PUBLISHER	SHORT FORM IN TEXT
Play, Dreams and Imitation in Childhood (Routledge & Kegan Paul Ltd, 1967 [W. W. Norton, 1951])	*Play, Dreams and Imitation*
The Child's Conception of Physical Causality (Humanities Press, 1966)	*Physical Causality*
Judgement and Reasoning in the Child (Humanities Press, 1947)	*Judgement and Reasoning*
The Child's Conception of Number (Humanities Press, 1952)	*Number*
The Child's Conception of the World (Humanities Press, 1929)	*World*
The Psychology of Intelligence (Routledge & Kegan Paul Ltd, 1967 [Harcourt Brace, 1950])	*Intelligence*
The Child's Conception of Space (Humanities Press, 1948)	*Space*
Logic and Psychology (Basic Books, 1957)	*Logic*
The Growth of Logical Thinking (Basic Books, 1958)	*Logical Thinking*
The Origin of Intelligence in the Child (Routledge & Kegan Paul Ltd, 1966 [W. W. Norton, 1963])	*The Origin of Intelligence*
The Construction of Reality in the Child (Basic Books, 1954)	*Reality*

The Moral Judgement of the Child (Routledge & Kegan Paul Ltd, 1950)	*Moral Judgement*
Piaget Rediscovered R. E. Ripple and V. N. Rockcastle (eds), (Cornell University Press, 1964)	*Piaget Rediscovered*
The Early Growth of Logic in the Child (Routledge & Kegan Paul Ltd, 1964)	*Early Growth of Logic*
Comments on Vygotsky's critical remarks concerning The Language and Thought of the Child, and Judgement and Reasoning in the Child (M.I.T. Press, 1962)	*Comments*

Other sources of reading in Piaget's psychology

The Psychology of the Child, by J. Piaget and B. Inhelder	(Basic Books, 1969)
The Child's Conception of Geometry, by J. Piaget, B. Inhelder, and A. Szeminska	(Basic Books, 1960)
The Mechanisms of Perception, by J. Piaget	(Basic Books, 1969)
The Child's Conception of Movement and Speed, by J. Piaget	(Basic Books, 1970)
The Child's Conception of Time, by J. Piaget	(Basic Books, 1970)
Mental Imagery in the Child by J. Piaget and B. Inhelder	(Basic Books, 1971)
An Outline of Piaget's Developmental Psychology for Students and Teachers, by R. M. Beard	(Basic Books, 1969)
A Teacher's Guide to Reading Piaget, by M. Brearley and E. Hitchfield	(Routledge & Kegan Paul Ltd, 1966)
The Developmental Psychology of Jean Piaget, by J. H. Flavell	(D. Van Nostrand Company, Inc, 1963)
The Growth of Basic Mathematical and Scientific Concepts in Children, by K. Lovell	(University of London Press Ltd, 1961)
The Pupil's Thinking, by E. A. Peel	(Oldbourne Press, 1968)

Section one

The process of intellectual development

This Section picks out some of the main threads which underlie Piaget's view of the nature of intellectual development. In making such a summary, much of the subtlety of Piaget's insights will be lost and some change of emphasis must result. Nevertheless, it is hoped that the following outline will serve as a guide for those who wish to read the psychology at first hand. At the outset, one point can be borne in mind. Piaget has divided the developmental sequence into stages and periods. The range of these is stated using chronological age. However, it is clear from Piaget's writings that the ages he gives for certain levels of thinking can be regarded only as guide-lines, or rough averages, for children's development. One can expect to find that there is considerable deviation from these norms. Some children do not reach the end of the developmental sequence. Some reach a given stage earlier or later than others. At any given stage in the sequence, modes of thinking characteristic of the earlier stages are present, and on occasions children may revert to modes of thinking which are more characteristic of earlier years.

The development of sensori-motor thinking: from birth to about two years

Piaget begins his analysis of the development of intelligence with a detailed study of the changes which occur in the baby's understanding of the world around him during the first two years of life. At birth, the baby has no awareness of self and not self, of an individual set in an environment. At that moment the world is spaceless, timeless and objectless, an undifferentiated experience of the present. However, the baby has a number of sensori-motor systems which can receive sensations arising from within his body and from the immediate surroundings to which he can make certain limited responses.

The starting point, or the 'given', of the developmental sequence, is these innate patterns of behaviour; like grasping, sucking, and gross body activity. It is in the interaction of these reflex-like patterns with the environment that a modification and development of behaviour occurs. This can be illustrated as follows. The tendency to suck can be exercised on any object which comes in contact with the lips, but the baby will soon become aware, through experience of sucking, that all objects do not have the same suckable properties. The lips and the mouth will register the shapes of objects, their size, hardness, softness or warmth. The alimentary canal will register the degree of nourishment such objects provide. There will be different sensations associated with sucking the nipple, the bottle, the thumb, or the corner of the pillow. By virtue of these experiences, the baby will register the differences between suckable objects, and the generalized tendency to suck will be modified by the kinds of objects available for sucking.

Changes occur, also, in other sensori-motor functions Vision is at first a reflex response to light intensity, but the eyes begin to focus on specific objects and to follow them as they move. Grasping is at first a reflex response to an object placed against the hand, but the hand begins to search, grasp and release objects without the initial tactile stimulus. Gradually these separate areas of reflex activity become co-ordinated. Having discovered by accident, for example, that the corner of the blanket is suckable, the baby becomes skilled at grasping it and bringing it to the mouth.

In summary, it can be seen that the internal needs of the baby are satisfied by the exercise of his reflex behaviour patterns upon the environment in which he is placed. The specific conditions in the environment cause a modification in these patterns. He makes simple discriminations (as between suckable objects, for example), and co-ordinates his separate behaviours in a rudimentary way (as between hand and mouth, for example). The result of this interaction of the infant with his surroundings is the acquisition of new behaviour patterns, or adaptations.

The co-ordination of sensori-motor actions continues, and the acquired behaviour patterns are applied to the environment in activities which are much more general than those from which such adaptations were first made. Any object which comes to hand, for example, is sucked to find its taste or hardness. Objects are banged to hear their sound. They are picked up, twisted, turned, and scrutinized. The baby is now applying acquired behaviours in order to maintain experiences which interest him. He may also try to prolong an interesting sight which is out of his control, by waving his hands or shaking his head. Another important development is seen at this time, which can be illustrated as follows. Say, for example, the baby has a teething ring suspended over his cot, and he has played with this by knocking it so that it swings from side to side. Now, suppose the ring is no longer hanging over the cot but hanging out of reach, and that the baby were to catch sight of it. He might respond to this sight by making brief pushing movements with his hands. Piaget deduces from this behaviour that the object—the teething ring in this case—exists in the baby's mind as a sensori-motor pattern which has been derived from the activity which the baby has performed with it. In other words, the recognition of the object occurs by means of a repetition of motor activity which has been performed with that object. At this stage in the baby's intellectual development, the mental action of recognition is not fully internal and produces a visible motor response which can be detected by the observer, i.e. the slight movement of the hand.

In summary, it can be said that the baby now begins to be active with his surroundings, rather than just upon himself. He strives to prolong those experiences which he has stumbled upon, by applying his patterns of adapted behaviour. He recognizes an object by recreating those actions he has performed with such objects.

The baby's response to hidden objects also changes at this time. Say, for example, that the mother plays a game with her baby in which she shows him a toy and then hides it behind her hand. Before about eight months, the baby would have shown no interest because the hidden object no longer existed for him. Now, however, he will push her hand aside and grasp the toy. This suggests that the baby is now beginning to dissociate objects from the actions he performs on them, and to this extent the world about him is acquiring some permanency and detachment from him. At about this time—eight to twelve months—the baby can anticipate events for he recognizes the implications of what is at that moment occurring; the door opening is a sign that someone will appear; mother putting on her hat is a sign that she will go away from him.

By the end of the first year exploratory behaviour, with all the parts of his surroundings within reach, becomes a dominant mode of behaviour. His actions become clearly intentional, and experimental. The noise made by a rattle has been discovered, and now, for example, he will further experiment with this experience by banging the rattle in a variety of ways to see what difference this makes. In the process of these experiments novel results will excite his interest. Say, for example, he drops the rattle on the pram cover just out of reach. Perhaps he stretches for it and inadvertently presses the cover down. The rattle slides towards him. This fortuitous method of attaining his goal will be followed up and he may repeat it several times. This discovery of new means through active experimentation, forms the basis of further changes. The attainment of his simple goals is, at first, very much a matter of trial and error. He applies all the procedures he knows until success is attained. However, the number of trials decreases, and success is attained without such apparent experimentation. This new development can be illustrated in the following example. Piaget describes the behaviour of a girl of sixteen months who is attempting to get a trinket out of a match box. The opening of the drawer of the box is too small for her to get her fingers in. She tries this, but fails. She then looks closely at the small opening, and Piaget notes that she opens her mouth wider and wider as if she were representing the opening of the box and the need to make it bigger. After this behaviour the baby pulls out the drawer of the box and reaches the trinket. Piaget suggests that the baby created a sensori-motor representation in the mind, of the relationships involved in this problem, and used this internal model for an experiment. Having discovered a means of solving the problem mentally, the baby then carries out the procedure in the environment. At this stage of the baby's intellectual development, the mental sensori-motor activity produced an external vestige which is visible to the observer, i.e. the movement of the mouth.

In this observation it can be seen that as the baby performs one activity in the environment his mental sensori-motor representations of earlier actions, which are reminiscent but not identical to his present environmental activity, begin to work, i.e. old sensori-motor representations operate in new situations, and this permits him to find a new way of behaving in the task at hand. The achievement of the ability to perform simple motor actions in the mind, and to apply them to environmental activity, marks a climax in the development of sensori-motor thinking. The baby has invented new methods of activity by means of mental combinations.

Another important achievement which appears at the same time, is that of the conservation of objects. The concept of an object can be defined as the realization, by the baby, that an object exists permanently and externally to him. The emergence of this conservation is gradual. At first, objects exist for the baby only in so far as they permit him to perform actions. The object is an extension of his actions. An object which is out of sight does not exist. Later in the developmental sequence, although the object has disappeared, he will continue his actions as if it were still present, and he will follow the path of moving objects with hand and eye. It is still the baby's actions, rather than the object, which endure. However, this endurance of motor actions leads the baby to follow the object into its hiding place. Once the search for hidden objects has begun, the fact that the object is out of sight divorces the action of searching from the object, and the object takes on an existence which is separate from the action it elicits. Piaget remarks:

'Henceforth the child places his own hand movements among those of external bodies, endowing the latter with an activity complementary to his own.' (*Reality*, p. 92.)

Thus objects become external and permanent, and a system of sensori-motor relationships is created between the child and the objects. These relationships are the representations with which the child carries out sensori-motor thinking.

The developments briefly described above, are referred to by Piaget, as the period of sensori-motor thinking and they extend to about the end of the second year.

SUMMARY

This period can be summarized as follows. At birth the infant has no knowledge of the existence of the world or of himself. His innate behaviour patterns are exercised in the environment and modified by the nature of the things he acts upon. His sensori-motor systems become co-ordinated during this activity. Gradually the baby builds up internal action models of the objects around him by virtue of the actions he has performed with them. He recognizes objects by means of these. This internal model of his actions allows him to perform mental experiments upon the objects he is manipulating physically. The result of performing actions with this internal model is sensori-motor thinking, i.e. internalized action.

The intellectual progress made during these two years is massive. Objects have become permanent, having an existence of their own, rather than being just an extension of the self. Events are related in that one experience presumes another and so cause and effect can be related. A temporal rhythm of events can be discerned, in which daily events follow each other in sequence.

This progress is also strictly limited. The child's understanding of the world does not go beyond those properties of objects and events which arise directly from his actions relating to them. He has a practical knowledge of the way things behave when he handles them, but no conception of why they behave as they do. His thought is locked in his own sensori-motor record, which is unique to him. His knowledge is private and not touched by the experience of others. The world of public knowledge embodied in the concepts conveyed through language can find no place in the model of the world he has so far elaborated.

RELEVANT SOURCES:

1. *The Origin of Intelligence in the Child.*
2. *The Construction of Reality in the Child.*

The emergence and development of symbolic thought: preconceptual representation: from about one and a half to five years

So far, it has been seen that all the activities the child has performed up to this time have created sensori-motor representations in the mind. The objects and events which form a part of these activities exist within the total mental model as replicas, or imitations, which derived from the actions performed with those objects, or within those events. It might be asked what these sensori-motor replicas are like. The answer to such a question must defy verbal description and can only be guesswork. Piaget's writings do suggest, however, a plausible way of looking at it. For example, the baby's representation of his mother might be as follows.

It will have been elaborated principally through all the activities he and his mother have shared. Feeding, bathing, bed-time routines, playing, going out together, will all have contributed to the image. These activities will have sensory connotations which will be built into the representation, like warmth and softness, the relief of hunger pains, a milky taste, a smell of soap, and soothing sounds. The representation will have spatial attributes, compounded from a tactile and visual exploration. Parts of the 'shape' will be clearly differentiated, others will not, depending upon the significance of their function towards him. The representation may have a sound linked to it, like 'ma-ma.' This representation will be of a most diffuse kind, for it will be linked with most of his other representations. The image of his mother which the baby has built up will occupy a predominant place in the child's sensori-motor scheme of things. She has satisfied his most compelling needs and his representation of her will have commensurate strength. How would a less significant aspect of his environment be represented? His representation of a plate, for example, would be a coalescence of all the activities he had performed with plates. It would be tied to other representations involved in the process of eating. Mixed with it would be a representation of mother, food, table, chair, spoon, cup, and so on. It would have, also, other sensori-motor properties related to its weight, shape, taste, sound, hardness and coldness.

These action replicas which exist in the mind at this time can be referred to as symbols of the experienced environment. It can be seen that the emergence of sensori-motor thinking, i.e. the use of mental representations to perform simple internal actions, can now be looked at from another point of view. The sensori-motor representation so far built up can be thought of as an interrelated collection of symbols which the child is able to manipulate in conjunction with his actions in the environment.

An example of the child's behaviour at this time will give clarity to this change of emphasis. Let it be imagined that a two-year-old is playing with his toys, scattered among which are some beads, a box lid, and a teddy bear. The following behaviour occurs. He places the beads upon the box lid and sets the teddy bear beside it. He then picks up the beads, one by one, and puts them to the mouth of the teddy bear.

This behaviour illustrates a number of important developments which the mental manipulation of symbols permits. The objects the child is using recall for him the sensori-motor representations he has had of the experience of eating. The box lid and the beads act as external repositories of his internal symbols of plates and food. The teddy bear symbolizes himself, perhaps. He organizes these objects as physical counterparts for his internal symbolization of the eating process, and in the course of this behaviour he achieves several things. By imitating all his earlier experiences of eating, he clarifies his mental representation of those events. Also, by playing in this way, his symbolization of the actual object he is using becomes refined. Most importantly, this procedure is exercising and developing the very process of symbolic mental activity. At this stage in his mental development, to stop him playing would be, more or less, to stop him thinking.*

This can be summarized as follows. The emergence of symbolic thought out of sensori-motor thinking, permits the child to:

1. Use his old sensori-motor representations in contexts which differ from those in which they were first acquired.
2. Use substitute objects in the environment to aid mental symbolic manipulation.
3. Divorce the representation of his own behaviour from his own body and apply it outside himself.

The manner in which the symbolic function develops from sensori-motor thinking can be summarized as follows. During sensori-motor thinking internal imitations of external activity have been constructed. By the end of the sensori-motor period the baby can recreate these imitations and so produce a mental image. The evocation of past activity in the present, Piaget refers to as deferred imitation. It is deferred imitations which produce mental images, and these mental images are the symbols which permit the further development of thought. Sensori-motor thought can be distinguished from symbolic thought in that in the former the internal imitation arises as a result of external activity, whereas in the latter deferred imitations or images arise first, and external activity follows from them.

* Much of the child's play during these years is concerned with the external manipulation of symbols which are heavily impregnated with his feelings about his social environment and his relationship to it. Although this activity is of great importance for his emotional and social development, no further reference will be made to this here, except to note that the mechanism for the expression of his feelings derives from the symbolic mode of thinking he is employing at this time.

SYMBOLIC THOUGHT AND LANGUAGE

At the time when symbolic thought emerges from sensori-motor thought, there is a dramatic increase in the child's use of language. Piaget attributes this fact to the emergence of the symbolic function, for words are themselves symbols. However, there is not a sudden change from the use of images, produced by deferred imitation, to the use of words as the basis of mental activity. At first, 'the word does little more than translate the organization of sensori-motor schemas to which it is not indispensable.' Language, at this time, is simply an accompaniment to action based on imagery. In the following examples, Piaget traces the change that occurs when the child begins to use verbal representation:

At 1: 11 (11) *(After she had been on a visit she said to me:) 'Robert cry, duck swim in lake, gone away.'

At 1: 11 (28) (When she was alone in the garden, she said to herself:)
'Mummy gone, Jacqueline gone with mummy.'

Piaget comments on these examples as follows:

> These behaviours are an illustration of the turning point at which language in process of construction ceases to be merely an accompaniment to an action in progress, and is used for the reconstitution of a past action, thus providing a beginning of representation. The word then begins to function as a sign, that is to say, it is no longer merely a part of the action, but evokes it. (*Play, Dreams and Imitation*, p. 222.)

Despite this spurt forward in the use of language it is Piaget's view that language, as a conceptual symbol system, is out of the child's reach at this time. Piaget's explanation of this might be summarized in the following way. On the one hand, there

* In these examples, and all those which follow, the age of the child is expressed as: At 1 : 11 (11), represents a child at the age of one year, eleven months and eleven days.

is the child's image-symbol system. These symbols are related one to another in a private and unique manner. Their relationships have arisen from the way in which the child has experienced objects and events, and, particularly, the actions he has performed in connection with them. He uses these symbols to signify the environment, and new experiences have meaning in so far as he can relate those experiences to his existent symbols. On the other hand, there is language which is a public symbol system, refined through centuries of usage. The words of which it is composed have socially agreed meanings, and relate one to another according to rules, or syntax. More particularly, words are used to embody concepts, and concepts, being abstractions, are not things or objects which can be seen or manipulated physically.

The conceptual property of words and word relationships with which Piaget is most concerned, at this stage, is the inclusion relationship. The inclusion relationship can be explained as follows. Words, like John Smith and Fido, label individual entities. Such entities have concrete existence. However, groups of these individual entities can be formed. For example, John Smith, Harry Jones, and Dick Brown, can be grouped and given the verbal label *man*. Fido, Rover, and Spot can all be labelled *dog*. These words, man and dog, are conceptual; they do not describe individual entities but groups of individual entities, and are therefore class words. Such words do not have concrete environmental counterparts; only exemplars of these concepts can be found in the environment. This process of conceptualization can be continued further. Take, for example, the classes man, woman and child. These can be grouped together and labelled *human being*. Alternatively, the classes man and dog might be grouped and labelled *animal*. The inclusion relationship arises because the individual entity, John Smith, is included in the class of men, and the class of men is included in the class of human beings, i.e. John Smith is a man, and a man is a human being.

These aspects of language as a symbol system, i.e. conceptual words, and their relationship to one another by inclusion, present great difficulty for the child when he uses language. The following imaginary example may illustrate this. Suppose the child has an image-symbol of his father. This has been elaborated from all the activities he has performed with his father or has seen his father perform. The image will also embrace men other than his father, like the postman, his uncle, and the man next door, for their shape and behaviour resembles that of his father. Thus, his symbol of his father is more like a vague impression of manly behaviour, for it has properties common to all men while at the same time symbolizing one man. Expressed abstractly, this symbol has both the properties of a general class, i.e. man, and of a specific exemplar, i.e. father.

Now the following imaginary incident might occur. Let it be assumed that the verbal symbol which the child has, is Daddy. The baker arrives at the door and the child says, 'Daddy', and mother replies, 'No! Not Daddy. He is a man.' The child repeats, 'Man', so giving his private symbol a public label. Later, his father returns and the child says, 'Man', to which his mother replies, 'No! Not Man. It is Daddy!' This same sort of confusion will occur in many other cases. For example, the child's symbol for a dog might be that one dog called Fido, which is the family pet, and inextricably mixed into it, one or two other dogs in the neighbourhood, big or small, plus the cows, horses, and cats he has seen. After all, these animals have striking similarities. What verbal label he gives to this symbol will depend upon whether he hears the word 'Fido' or 'dog', when he has the symbol in mind, and this labelling will have no connection with the conceptual nature of those words, or their relationship of inclusion. As far as the child is concerned, a Fido could be a dog or a dog could be a Fido, with equal logic.

The child's images act, therefore, both as symbols for an

individual identity and as symbols for a class of entities. Whereas in language, different words are used for the exemplar of a class and for the class itself, the different words being connected by the inclusion relationship. Commenting on the child's imagery, Piaget remarks:

> These two characteristics—absence of individual identity and of general class—are in reality one and the same. It is because a stable general class does not exist that the individual elements, not being assembled within the framework of a real whole, partake directly of one another without permanent individuality, and it is the lack of individuality in the parts which prevents the whole from becoming an inclusive class. (*Play, Dreams and Imitation,* p. 226.)

The child's symbolic activity at this time, Piaget calls preconceptual, by which he implies that the symbols available for mental manipulation, and which are expressed in language, have the property of a preconcept. A preconcept is 'intermediary between the imaged symbol and the concept proper', and is defined as: '. . . absence of inclusion of the elements in a whole, and direct identification of the partial elements one with another, without the intermediary of the whole'. (*Play, Dreams and Imitation,* pp. 226–7.)

The following are examples which illustrate the child's preconceptual thinking:

At 2: 6 (3) 'That's not a bee, it's a bumble bee. Is it an animal?'

At 3: 2 (20) (Child) 'Is that man a daddy?'
(Adult) 'What is a daddy?'
'It's a man. He has lots of Luciennes and lots of Jacquelines.'
'What are Luciennes?'
'They're little girls and Jacquelines are big girls.'

At 3:3 (27) 'Are little worms animals?'
(*Play, Dreams and Imitation,* p. 225.)

In these examples, the inclusion relationship between a bumble bee, a bee, and an animal, or between a worm and and an animal, is not yet understood. The defining attribute of a daddy, as a class of man having children, is understood, but expressed using exemplars rather than the conceptual word 'children'. The concepts 'little girls' and 'big girls', are expressed by reference to individual exemplars.

Clearly, the child's use of language at this time will play a part in the development of his mental processes. Piaget sees the progress of this development as one in which the private image gives way to the public verbal sign. When operational thinking emerges, at around seven or eight years, the verbal sign is the signifier used in thought, rather than the image. The image then acts as an individual symbol which supports the verbal sign. Commenting on this metamorphosis he writes that at the level of operational thought, the image '. . . . is reduced to the rank of mere symbol, inadequate though sometimes useful, its role being that of a mere assistant to the verbal sign'. (*Play, Dreams and Imitation,* p. 271.)

From the point of view of the relative importance of language to the progress of intellectual development, Piaget considers that there are a number of factors at work, one of which is language. Piaget's main concern is to discover the operational systems by which thought is organized, and he regards language, and its use by the child, as a factor which contributes to the formation of these operational systems. He comments as follows: 'Obviously, since conceptual schemas are related to the system of organized verbal signs, progress in conceptual representation will go hand in hand with that of language.' (*Play, Dreams and Imitation,* p. 221.)

However, Piaget points out that to use language in a conceptual manner, and as a vehicle for thinking, requires the existence of operational systems. He comments that, '. . . the capacity for constructing conceptual representations is one of the conditions necessary for the acquisition of language'. (*Play, Dreams and Imitation,* p. 221.)

Thus, for Piaget, there is a reciprocal process in which the child's use of language helps to develop his mental operations, while, on the other hand, the development of operational thought, due to the working of many factors, permits language to be used in operational activity. Writing about the growth of logical thinking in early childhood, Piaget says:

> . . . the fact that the language of adults crystallizes an operational schema does not mean that the operation is assimilated along with the linguistic forms. Before children can understand the implicit operation and apply it, they must carry out a structurization, or even a number of successive restructurizations. These depend on logical mechanisms. They are not passively transmitted by language. They demand an active construction on the part of the subject. (*Early Growth of Logic,* p. 4.)

At the end of the book, under the heading, 'Problems and Difficulties', a brief mention is made of the fact that not all psychologists would accept this view of the role played by language in the development of thought.

TRANSDUCTION

The following examples illustrate a prevalent form of thought during these years, which Piaget refers to as transduction:

At 2: 4 (27) 'Daddy's getting hot water, so he's going to shave.'

At 4: 10 (21) 'I haven't had my nap, so it isn't afternoon.'
(*Play, Dreams and Imitation,* pp. 231–2.)

In these examples, the children are making statements of implication, i.e. if x, then y, although there is not necessarily any relationship between the two events. Hot water implies shaving, and no nap implies no afternoon. Piaget suggests that reasoning of this kind occurs because the child is attempting to make inferences without having the concepts with which to carry out the reasoning process. Reasoning, of the kind used here, needs a general symbol (concept) of shaving, within which there are a number of exemplars, e.g. shaving with hot water, shaving with an electric razor, shaving with brushless cream. A general symbol (concept) for hot water would also be required, within which there are exemplars, e.g. hot water for shaving, hot water for making tea, and hot water for washing. Since the child has only preconcepts with which to reason at this stage, i.e. symbols which are neither general nor particular, transductive reasoning results. It might be imagined that the child's symbol of shaving has in it, Daddy, face, hot water, razor, soap, bathroom, and the symbol for hot water has in it, washing, face, soap, bathroom. Thus, the inference that hot water implies shaving is ready made. (Hot water is shaving.) Transduction arises when the child reasons from preconcept to preconcept, and it is tempting to refer to this process as illogical. However, one can hardly think illogically before one can think logically. Piaget would use the term prelogical to mark out this distinction.

JUXTAPOSITION AND SYNCRETISM

These terms refer to modes of thinking which are closely related to the nature of the symbols the child has so far elaborated (preconcepts), and the form of reasoning which preconcepts permit (transduction.) Juxtaposition and syncretism express themselves in the way the child explains the behaviour of things (cause and effect), the way he expresses his thoughts verbally (sentence structure), and the way he pictures his understanding (drawings.) To juxtapose is to collect together parts without relating them:

At 4: 0 (Adult) 'What makes the engine go?'
'The smoke.'
'What smoke?'
'The smoke from the funnel.'
(*Physical Causality*, p. 228.)

Here, the explanation is derived from what is visible without a knowledge of the working parts. The movement of the engine and of the smoke, are both obvious and dynamic, and are juxtaposed as cause and effect.

At 6: 0: 'I've lost my pen because I'm not writing.'
(*Judgement and Reasoning*, p. 17.)

Here, the difficulty in the use of words of implication is seen.

The drawing below, by a backward nine-year-old, shows juxtaposition without relations in the case of the parts of a bicycle.

Piaget remarked about this kind of drawing that the child knew 'that the chain, the pedals and the cog-wheel were necessary to set the machine in motion', but he did not have a 'knowledge of the details of insertion and contact'.

Ru

(*from Physical Causality*, p. 207.)

On the other hand, syncretic thought is seen when the child relates everything to everything else.

At 6: o (Adult) 'Why does the sun not fall down?'
'Because it is hot. The sun stops there.'
'How?'
'Because it is yellow.'
(*Judgement and Reasoning*, p. 229.)

At 4: o (Adult) 'How does the bicycle go?'
'With wheels.'
'And the wheels?'
'They are round.'
'How do they turn?'
'It's the bicycle that makes them turn.'
(*Physical Causality*, p. 201.)

In summary, it can be said that juxtaposition is thinking which arises through concentration upon the parts or details of an experience without relating those parts into a whole, while syncretism is thinking which arises through concentration on the whole of an experience without relating the whole to the parts. Both these modes of thought arise through the inability to synthetize parts and the whole into a related group. Piaget remarks about this as follows:

For the child, 'everything is connected with everything else, which comes to exactly the same thing as that nothing is connected with anything else'. (*Judgement and Reasoning*, p. 61.)

CENTRATION AND STATIC REPRESENTATION

Piaget gives many examples of the forms of thinking which he refers to as static, and showing centration. The following example illustrates these:

At 4: 3 An experiment with eggs and egg-cups.

Arrangement A

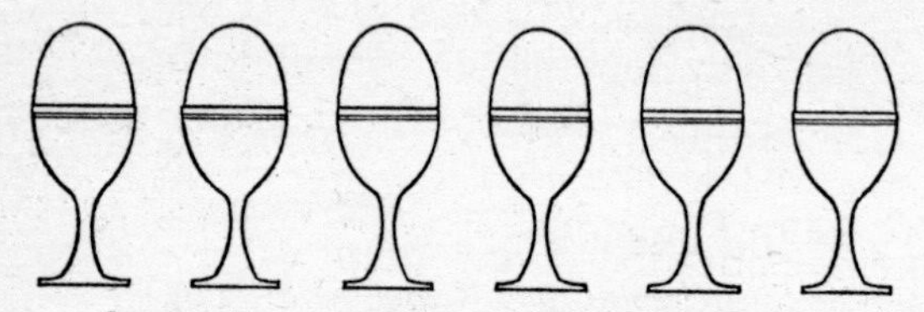

A row of eggs in egg-cups is arranged as shown. The child is asked if there is the same number of eggs as egg-cups, to which he replies, 'Yes.'

Arrangement B

(Adult) 'Are they the same now?'
'No.'
'Why?'
'There are more egg-cups.'

Arrangement C

(Adult) 'Look, now is there the same number of eggs and egg-cups?'
'No, there are more eggs.'

(*Number*, pp. 50–1.)

Piaget explains thinking of this kind in the following way. As has been seen above, children of this age are not able to represent a complex object or event, relating parts to form a whole. They may either experience a total undifferentiated whole (syncretism), or concentrate upon parts or details (juxtaposition). The replies of the child in this experiment indicate that he is either attending to the arrangement of the eggs or to the arrangement of the egg-cups, but not to both and their relationship to each other. When the egg-cups are spread apart, they appear to have increased in number. When the eggs are spread apart, they, too, dominate his thinking and appear to have increased in number. This fixation upon one aspect of a changing relationship to the exclusion of other aspects is referred to by Piaget as 'centration'.

Closely associated with this inability to decentre thought is the child's inability to link one arrangement of objects with another as he watches such an arrangement change. Had the child been able to relate Arrangement A with Arrangement B, and Arrangement B with Arrangement C, then he would have known that each arrangement was a transformation of the other. In other words, he would have known that the parts of the experience had not changed, only the arrangement of those parts. This inability to manipulate mental representations in a rapid and flexible way, and so be able to handle transformations, is referred to by Piaget as static representation.

EGOCENTRISM

Piaget defines the term egocentrism as follows: '... egocentrism is on the one hand primacy of self-satisfaction over objective recognition ... and, on the other, distortion of reality to satisfy the activity and point of view of the individual. In both cases it is unconscious, being essentially the result of failure to distinguish between the subjective and the objective.' (*Play, Dreams and Imitation,* p. 285.)

The following examples illustrate some of the forms egocentrism may take:

At 3:7 (14) 'The stairs are horrid, they hit me.'
(*Play, Dreams and Imitation*, p. 252.)

At 6:0 (Of the sun) 'It comes with us to look at us.'
'Why does it look at us?'
'It looks to see if we are good.'
(*World*, p. 216.)

At 3:2 (20) Lucienne heard a waggon on the road at right angles to the one on which we were, and was frightened:
'I don't want it to come here. I want it to go over there.'
The waggon went by, as she wished.
'You see it's gone over there because I didn't want it to come here.'
(*Play, Dreams and Imitation*, p. 258.)

At 4:6 'I'm stamping, because if I don't the soup isn't good enough. If I do, the soup's good.'
(Ibid.)

At 6:0 (Adult) 'Why are there waves on the lake?'
'Because they've been put there.'
(*Physical Causality*, p. 89.)

At 4:6 (Adult) 'Have you got a sister?'
'Yes.'
'And has she got a sister?'
'No, she hasn't got a sister. I am my sister.'
(*Judgement and Reasoning*, p. 85.)

Underlying all these expressions of egocentrism, it can be seen that the common factor is the subjective and affective nature of the child's view of the world. He credits the inanimate with feelings like his own. He believes his thoughts have the power to change events. (After all, his thoughts and his actions are two parts of the same process. His actions change events, so why not his thoughts?) He believes things exist because someone has put them there, rather as his mother provides things for him. Lastly, he does not yet have any notion of viewpoints different from his own.

SUMMARY

Symbolic representation emerges from sensori-motor representation at about the two-year mark. Sensori-motor thinking continues in parallel with this symbolic activity. The symbolic function arises because internalized imitation—the end-product of sensori-motor thought—can be evoked in the absence of the actions which originally created the imitations. This evocation Piaget refers to as deferred imitation. Deferred imitations give rise to images, which are the symbols that the child uses for preconceptual thinking. His image-symbols are a collection of actions, objects and events, which are related to one another in a unique and private way.

By virtue of the symbolic function, the use of language becomes possible. The two symbol systems, the child's collection of images, and language, are not mutually supporting at the start. Language is non-representative and conceptual in its organization, whereas the child's symbols are closely related to their sensori-motor origins. The preconcept develops, which is 'intermediary between the imaged symbol and the concept proper'. Preconcepts are representations which achieve neither true generality nor true individuality but fluctuate incessantly between the two extremes.

The uniquely personal nature of the child's representations

during these years restricts his use of them in a social context, i.e. when they are expressed in language. When he does not respond to the thoughts of others, this is due in part to his inability to fit such thoughts into his own. What might be loosely called his selfishness is, in part, a measure of his inability to think about any other point of view except his own. Preconceptual thought shows properties like transduction, juxtaposition, syncretism, centration, static representation and egocentrism.

No mention has been made of the child's understanding of space and time during these years. As far as the representation of space is concerned, his understanding of it is limited to the sensori-motor record of the physical activity he has performed when moving about among the concrete objects of his environment, and in handling those objects. Space itself, an invisible and intangible concept, has no existence. For him, space is embodied in the shape of things, in nearness, separation, enclosure and continuity. He cannot represent groups of objects in any other way than the one in which he sees them at a given moment. He will recognize an object from an unusual point of view only if he has a complete sensori-motor record of it. Distant groups of objects or scenes will not have this permanence of form, and will be regarded as new and different when seen from an unusual viewpoint.

Time, also, as an invisible and intangible concept, has no meaning for him. Nevertheless, several kinds of experience have temporal connotations. He recognizes the rhythm of his daily life. Meals, play, sleep, light and dark, his father's presence or absence in the house, are related in sequence. The duration of an event will exist by virtue of the sensori-motor activity he performs in order to follow movement.

RELEVANT SOURCES

1. *Play, Dreams and Imitation in Childhood.*

2. *The Child's Conception of Causality.*
3. *Judgement and Reasoning in the Child.*
4. *The Child's Conception of Number.*

Articulated or intuitional representation, the threshold of operational thinking: from four to eight years

The properties of preconceptual thought which have been outlined above show changes during these years. A metamorphosis takes place, and operational thought emerges. This transition can be viewed either from the point of view of the gradual disappearance of the limitations of preconceptual thinking, or from the point of view of the emergence of operational thinking.

The purpose of this sub-section will be to indicate the way preconceptual thinking becomes articulated, showing an intuition of operational structure, thus overcoming the limitations of the preconcept. The next sub-section will indicate the way operational thinking emerges from this intuitional stage.

SOCIAL INTERACTION AND LANGUAGE

The child's increasing social involvement during these years gives impetus to the development of his intellectual processes. Social interaction requires communication and the child attempts to express his thoughts and to make sense of the thoughts of others. Such communication is difficult, for to some extent he, and his peers, are still living in unique and private worlds which do not lend themselves to interchange and reciprocity. However, the sharing of materials, the sharing of experience in play, and the engagement in similar tasks, force upon the child a communal form of thought. The principal currency of his social interchange is language, and he is immersed in a sea of words which define and relate his social behaviours and his physical activities. Whether he likes it or not, the child begins to see his relationship to others as reciprocal and not unidirectional. He discovers that what he thinks is not necessarily the same as the thoughts of others. Social activity and the linguistic framework within which it operates press upon him, and he adjusts his thoughts in conformity. He begins to see himself and the world around him, from other points of view. Piaget says: 'In fact, it is precisely by a constant interchange of thought with others that we are able to decentralize ourselves in this way, to co-ordinate internally relations deriving from different viewpoints.' *Intelligence*, p. 164.)

During these years, the child's symbols begin to relate to each other as the words relate in speech patterns. Language begins to operate as a vehicle for thought. One is reminded in this connection of an earlier stage in the development of thought. At the time when symbolic thought emerged from sensori-motor thought, the baby had created internal images of his overt activity, and he used these imitations in order to carry out mental actions. Now speech is becoming internalized and can play a greater part in mental actions.

THE DECLINE OF THE PRECONCEPT

At 4: 4 (2) (The child is looking at an iron bar.) 'What's that stick, is it iron?'
'Yes.'
'Oh, yes, because it's cold, because it makes music.' (Hitting the ground with it.)
(*Play, Dreams and Imitation*, p. 233.)

This example shows that a primitive concept of an iron stick has formed, defined by its coldness and the sound it makes. This classification differentiates it from, say, a wooden stick.

THE DECLINE OF TRANSDUCTION

At 6: 7 (8) (Child) 'Do blue butterflies like the wet?'

'Yes'.
'And the brown ones? They like it to be dry.—
Then why are there some here with the blue ones?'
(Ibid., p. 232.)

In this example, the child reaches a false conclusion about the brown butterflies liking it to be dry. Nevertheless, the transductive element is not so strong, for the child notes that the evidence does not support the reasoning.

THE DECLINE OF JUXTAPOSITION AND SYNCRETISM

At 8:0 (Child) Discussing steam engine: 'There's a big fire. The fire makes a bit of iron go, that's sort of bent (connecting rods), and that makes the wheels turn.'
(*Physical Causality*, p. 229.)

At 6:0 (Child) The bicycle goes with the wheels and 'the gentleman makes them work'.
'How?'
'When he's riding. He pedals with his feet. That makes the wheels work.'
'What is the chain for?'
'To hold the wheels, no the pedals . . .'
(Ibid., p. 207.)

These examples indicate that the details and the total process are now seen much more as a related whole, and there is more understanding of cause and effect. As far as the child's drawings are concerned, they also indicate a parallel development.

THE DECLINE OF CENTRATION AND STATIC REPRESENTATION

At 5:7 (The experiment with the eggs and egg-cups.)

Arrangement A

The child puts six eggs in six egg-cups. They are then rearranged as follows:
Arrangement B

(Adult) 'Where are there more?'
'Here.' (Indicating the eggs.)
'If we wanted to put one egg back in each egg-cup would there still be the right number?'
'Yes . . . I don't know'.

(*Number*, p. 52.)

The tendency to fix upon one aspect of the arrangement is declining. The child begins to decentre, and to be able to relate one arrangement with another. He begins to think in terms of a transformation rather than separate unconnected states. This ability has not yet fully emerged in this example.

THE DECLINE OF EGOCENTRISM

At 5: 11 'You'd think the stars are moving, because we're walking.'

At 6: 3 (10) (J. was whirling round.) 'You can feel it going round, but things aren't really turning.'

(*Play, Dreams and Imitation*, pp. 265 & 260 respectively.)

The child is no longer sure that the stars move because he moves. Whirling round only seems to make the world revolve.

As far as spatial relationships are concerned, his decentralization is still some time ahead as this experiment shows.

At 8 : 2 The child sits before a model of three mountains in the position shown in the diagram. A doll is placed as indicated. From a set of pictures, which are views of these mountains from different perspectives, the child is asked to choose the one which he thinks is the doll's view.

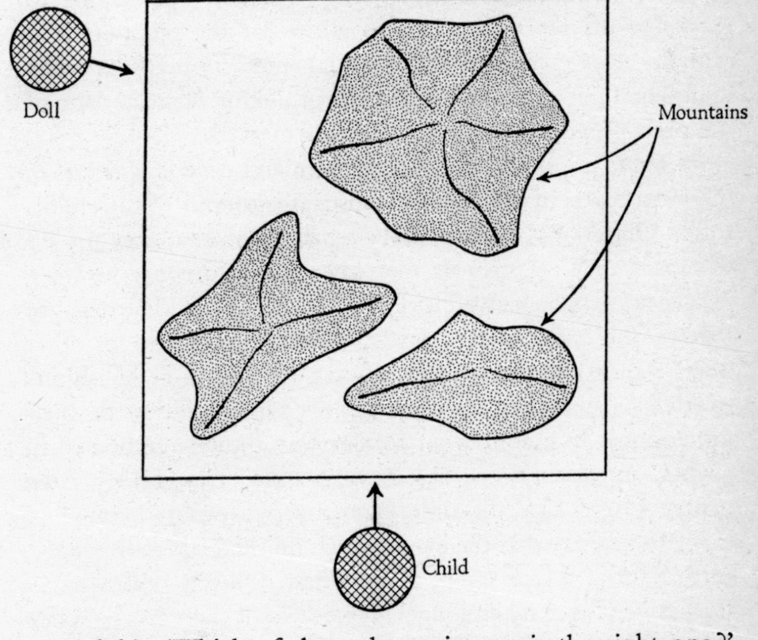

(Adult) 'Which of these three pictures is the right one?'
'This is the right one; he sees the three mountains just as they are.'

(The child chooses a picture of his own view.)

(*Space*, p. 219.)

SUMMARY

Social interaction with language, makes an important contribution to the development of mental structures between four and eight years of age, and continues to do so from now on. These factors are influential in decentralizing the child's view of the world. The more the child becomes socially related and uses language in his activity, the more he reorientates his mental model of the environment. The change occurs in two main ways. Firstly, he orders and relates his representations more in accord with the conceptual nature of language. This in turn increases his ability to communicate coherently. Secondly, he begins to rearrange his representations to allow for the relativity and plurality of viewpoints which social interaction forces upon him. The limitations in the forms of thinking characteristic of the preconceptual period become less marked.

As far as an understanding of space and time is concerned, these concepts, in the abstract sense, are still out of the child's reach. Objects have space, and he may think that they use up the space they occupy. He may say that the distance between people is less when a wall is placed between them. The distance between objects is not constant because the presence of other objects may change such distances. The position of objects relative to him may also change his representation of their dimensions. In a somewhat similar way, time is embodied in events, and each event has its own time. The child will be unable to compare the time taken by two events because of this. He may say that the event which finished last took longer, regardless of which event started first. Objects which move farther may have taken longer, regardless of their speed. When things are large, they may be thought to be old also. He may think the time of day in adjacent places to be different.

Towards the end of this age range, children begin to understand that space can be empty as well as full, and that non-events, i.e. the interval between events, have duration.

At 5 : 9 (2) 'Are there times when there aren't any hours, or are there always, always hours?'
(*Play, Dreams and Imitation,* p. 266.)

Piaget explains the child's difficulty with spatial and temporal measurement in the following way. Not only are space and time invisible, but also they are both totally interrelated as far as the child is concerned. Time is used up in movement, and movement uses up space. Space is appreciated by movement in it, and movement has duration.

The disappearance of the main characteristics of preconceptual thought can be looked at in another way. It can be said that the mental actions producing representations are becoming more flexible and mobile and co-ordinated, one with another. There is a growing mobility by which the child can group his representations into an interrelated system. This co-ordination of representations is the indication that mental operations are beginning to emerge. The nature of operational thinking will be outlined in the next sub-section. The forms of thought which begin with the emergence of symbolic representation, at around two years of age, and develop into the articulated representations at around seven or eight years of age, constitute the period of Pre-operational Thinking.

RELEVANT SOURCES

1. *Play, Dreams and Imitation in Childhood.*
2. *The Child's Conception of Causality.*
3. *Judgement and Reasoning in the Child.*
4. *The Child's Conception of Number.*
5. *The Child's Conception of the World.*

The emergence of operational thinking: concrete operations: from seven to twelve years

The mental processes, which Piaget refers to as operations,

emerge from the articulated representations which have been outlined above. The purpose of this sub-section is to pick out the main properties of concrete operations and to show how they have formed from the earlier modes of thinking. A comparison will be made between pre-operational and operational thinking.

OPERATIONS WITH CLASSES

At 6: 8 (Pre-operational.) The child is given a set of wooden beads, eighteen of which are brown and two of which are white.

(Adult) 'Are there more wooden beads or more brown beads?'

'More brown ones, because there are two white ones.'

'Are the white ones made of wood?'

'Yes.'

'And the brown ones?'

'Yes.'

'Then are there more brown ones or more wooden ones?'

'More brown ones.'

(*Number*, p. 164.)

CO In order to answer correctly the questions about the brown and white wooden beads, the child must perform certain mental actions with three classes of beads. He must put together mentally the class of white beads and the class of brown beads to form the inclusive class of wooden beads. He must also be able to reverse this operation mentally, so that he separates the class of white beads from the wooden beads, to re-form the class of brown beads. At the same time as he does this, he needs also to maintain the class of wooden beads so that the class of brown beads can be included in it. The answers the child gives show that he cannot perform these mental operations. He does not compare part (brown beads) with whole (wooden beads), but instead he compares part (brown beads) with part (white beads). When a part is removed, the whole no longer exists.

At 8 : 0 (Operational.)
(Adult) 'Are there more brown beads or more wooden beads?'
'More wooden ones.'
'Why?'
'Because the two white ones are made of wood as well.'
(*Number*, p. 176.)

The answers the child gives here indicate that he finds no difficulty with this experiment. He is able to reverse the mental action which created an inclusive class of wooden beads, to re-form the classes of white and brown beads, yet at the same time the inclusive class of wooden beads is maintained. The child then compares part with whole. The mental operation the child employs in this experiment, Piaget refers to as reversibility.

Many younger children are able to sort objects into groups or find names for groups of objects. But this does not imply that they are able to operate upon their representations. As was seen earlier when outlining Piaget's view of mental activity with the preconcept, children may often use conceptual words but they do not have the logical property of a concept. Piaget remarks:

> A number of writers found that children of 2–4 would tell them that a dog was an animal, a lady was a person and a daisy was a flower. They concluded that these children had reached the level of hierarchical classification. To this we cannot agree. What these facts indicate is that, given certain familiar elements, these tiny children can reach beyond the level of graphic collections, and the corresponding linguistic

schemata are structured into parts and wholes. But the structure is not that of an operational classification. (*Early Growth of Logic*, p. 117.)

OPERATIONS WITH RELATIONS

At 7:0 (Pre-operational.) The child is given two different shaped glasses, a wide glass (B), in which there is some liquid, and a narrow glass (A), which is empty. The child is asked to pour liquid into (A), so that there is the same amount of liquid as there is in (B). He does this as shown.

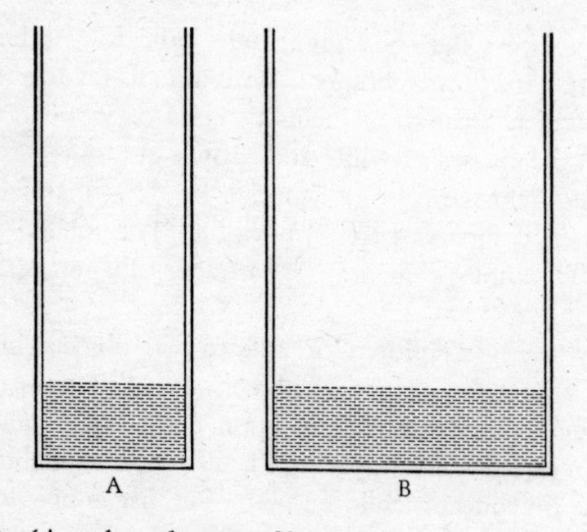

(Adult) 'Are they the same?'

'No.'

The child then adds more liquid to (A), as shown below:

The child then says, 'No, it's too much', and then he pours out the liquid to make the levels equal again. (*Number*, p. 16.)

In order to carry out this experiment successfully, the child must be able to transform the shape of the liquid he sees in

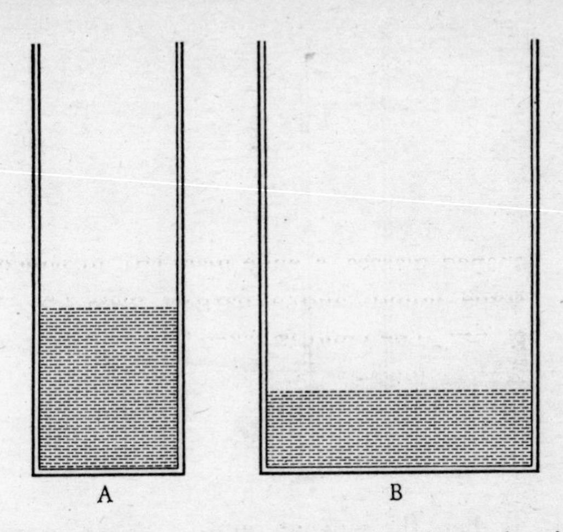

(B) into a new form, which will be the shape that is created as he pours the liquid into (A). To express this in another way, the child must have the mental operation which permits him to be aware that the amount of liquid in (A) can be made equal to that in (B), if the difference in the widths of the two vessels is compensated by the difference in the heights of the liquids. This mental action, Piaget refers to as 'equating the differences'.

In the above experiment, Piaget explains the child's answers as follows. The child pours the liquid into (A) to give equal heights. He is centring on height and ignoring the width. When asked if there is the same amount, he looks again and says, 'No', because he is now centring on width and ignoring the height. He then pours in more liquid, looks again and says, 'No, it's too much', because his centration has changed again to height. He therefore pours out the liquid to give equal heights again.

At 6: 6 (Operational.) The same experiment. The child fills (A), as shown, giving nearly equal amounts of liquid.

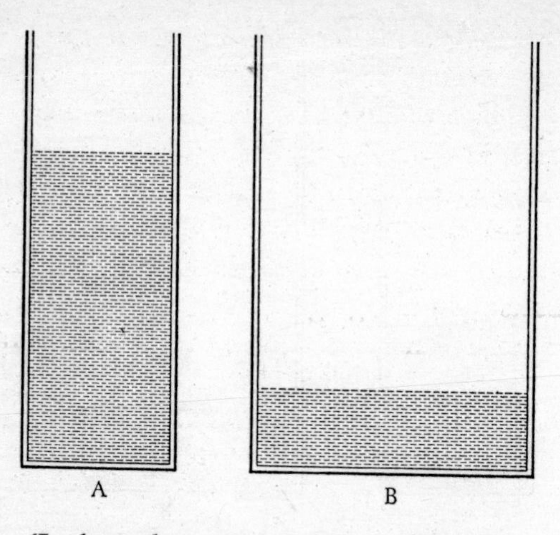

(Adult) 'Is there the same amount to drink?'
'Yes, it's the same.'
'Why?'
'Because it's narrower here (A) and wider here (B).'
(*Number*, p. 19.)

The mental operation which underlies the ability to understand this problem, is again the property of reversibility. However, the form of the reversibility is of a different kind to that seen in the operations with classes. Reversibility with classes is achieved by performing an opposite action which will undo the first action, e.g. taking apart as opposed to putting together. The reversibility of relations, on the other hand, is achieved by performing a second action, which exactly compensates for the first condition without undoing it. The result of the two conditions, together, produces an equivalence.

CONSERVATION

In the experiment above, with the liquids, the child who solves the problem has achieved what Piaget refers to as the conservation of quantity. That is to say, the child can express with certainty that the amount of liquid can be the same, regardless of differences in its shape. The conservation of quantity, or substance, is one of several conservations which the child achieves through operational thought. The other conservations include number, class, length, weight, area, and volume. In the general sense, conservation could be defined as an operational process of the mind which produces the realization that certain aspects of a changing condition are invariant, despite those changes. It can be seen that conservation and reversibility are closely related, and Piaget expresses this fact as follows: 'Conservation has thus to be conceived as the resultant of operational reversibility.' (*Logic,* p. 9.)

The child's acquisition of the various conservations begins at around six to seven years (except in the case of the conservation of objects, which appeared at the end of the sensorimotor period.) Substance (quantity) may be conserved between six to eight years, weight at around nine or ten years, and volume at, perhaps, eleven or twelve years. However, there is a wide age range at which children acquire the various conservations. As far as the order of acquisition of the various conservations is concerned, Piaget makes the following comment, '. . . it is more difficult to order serially, to equalize, etc., objects whose properties are less easy to dissociate from one's own action, such as weight, than to apply the same operations to properties which can be objectified more readily, such as length.' (*Logical Thinking,* p. 249). It can be noted also that some conservations rest upon earlier conservations, for example velocity, and therefore will appear later rather than earlier.

THE NATURE OF CONCRETE OPERATIONS

The two experiments outlined above permit a glimpse of the kind of mental activity which Piaget would call operational. In order to define the precise nature of operational thought, Piaget has cast it into mathematical form. It is not the intention to pursue the intricacies of this here, because it will cloud the issue as far as the purpose of this Section is concerned. The ensuing analysis of the main properties of operational thought, at this stage of the child's development, will therefore avoid recourse to the mathematics of it. Some mention of this aspect will be made in the next Section.

The essence of a mental operation, which distinguishes it from the mental activity seen earlier, is the way in which representations are organized by the mind. In other words, the system by which one representation is related with another. Thus, to describe an operation is to describe not so much its content, but rather the way it behaves. Any discussion on the nature of operations will devolve, therefore, upon the rules or laws which seem to govern the way in which mental activity occurs during these years. (See pp. 82 and 94.)

A most significant aspect of Piaget's concept of a mental operation is that it is seen as an action performed by the mind, or, more precisely, a set of related actions which form an integrated whole. Thus, a mental operation has not one property, but a group of properties, each of which depends upon, and is necessary for, each of the other properties. Piaget conceives of these interrelated properties as having evolved from the internalization of physical actions first performed in the environment. It is to be expected, therefore, that such mental actions will show properties in some ways similar to physical activity. These similarities can be illustrated as follows. For example, separate motor actions can be combined together to form a new action. This occurs when the hand and eye are co-ordinated in some purpose, or when separate objects are put together to form new objects. Physical actions of this kind can be performed in varied ways, yet still achieve the same result, as, for example, when a destination is reached taking one of several alternative pathways, or when a tin is filled with beads by putting them in one by one, or in one handful. Thus, an operation is a mental action in which representations are combined to form new representations and in which such combinations can be achieved in various ways.

A second important property, which is the hallmark of an operation, is that it is reversible. The property of reversibility expresses itself in two ways, both of which have physical counterparts. The physical counterpart of reversibility, as it occurs with classes, is seen, for example, when a tin full of beads is emptied after it has been filled, or when a movement in space is retracted. This form of reversibility, the combining of representations, followed by their separation, is called inversion. (Alternatively—negation or elimination.) The physical counterpart of the second form of reversibility, as it occurs with relations, is seen, for example, when the hand holds an object steady by applying a force equal and opposite to the weight of the object, or when the eyes move to compensate for the changes in the position of the head and so keep an object in vision. This form of reversibility, the translation of relationships into equivalent forms, is called reciprocity. (Alternatively—symmetry or equivalence.) Thus an operation is a mental action which exhibits reversibility. This reversibility expresses itself in two forms, those of inversion and reciprocity.

The emergence of mental operations is an important step forward in the development of the child's thinking. However, there are a number of limitations to this newly won autonomy. The mental actions by which classes and relations are formed bear largely upon a perceptual environment. This environment can be classified in a number of ways and equivalences can be formed, but the activity is not far removed from rearranging mentally that which could also be rearranged physically. If, for example, the child is presented with a problem expressed in verbal form without any physical counterpart, he may be unable to perform the mental actions which are required to find the solution to the problem. On the other hand, if he were given the objects to manipulate, the problem might be solved. Piaget makes this important limitation clear by calling these mental actions concrete operations. He comments as follows, '. . . concrete operations consist of nothing more than a direct organization of immediately given data'. (*Logical Thinking,* p. 249.)

The operations the child can now perform are also capable of application beyond the 'immediately given data', but this potential is not realized at first. For example, if the child has arranged a set of regular shapes in order of increasing number of sides, he will be aware that further shapes could be added at one end, each with one more side than its predecessor. It is not likely, however, that the child would be aware that if this process were extended indefinitely, the result would be a circular shape. In the case of classification, he will be aware that cars travel on land, ships on the sea, and aeroplanes in the air. He might also reach the conclusion that there could be 'ships which go under the sea'. But this realization would not extend to describing what these 'ships' would be like, unless he had some direct or indirect experience of them. Piaget comments on this as follows:

> In sum, concrete thought remains essentially attached to empirical reality. The system of concrete operations—the final equilibrium attained by pre-operational thought—can handle only a limited set of potential transformations. Therefore, it attains no more than a concept of 'what is possible', which is a simple (and not very great) extension of the empirical situation. (*Logical Thinking,* p. 250.)

These limitations do not imply that operations cannot be performed on past experience or upon a world of fantasy, as, for example, when old experiences are recreated in play, or when thinking is concerned with fairies, gnomes or angels. In such cases, these mental representations may be subject to the modes of thinking which are operating in this period. A further important limitation occurs in the restricted nature of the acts of conservation. The various conservations do not appear in all areas at once, but occur sequentially in more diverse areas of the child's activity. It can be noted, also, that the conservation of a specific invariant, like substance for example, may not become generalized in all instances of that invariant at the time of its first appearance. The child may conserve the quantity in a lump of plasticine when it is rolled out into a short sausage, but when it is rolled out into a longer one, he may falter in his belief.

As far as the reversibility of mental actions is concerned, the period of concrete operations has an important restriction. The two forms of reversibility—inversion and reciprocity—remain as separate operational systems. Thus, while the child can perform an operation with a class by inversion, or operate on a relation to form an equivalence, he cannot utilize both these forms of reversibility at the same time. Piaget comments on this as follows: 'No groupments are present at the level of concrete operations to combine these two kinds of reversibility into a single system.' (*Logic,* p. 29.)

Nevertheless, a most important benefit accrues from the emergence of concrete operations. The child is now able, potentially, to operate with the symbol systems of language and mathematics. For example, he can organize words conceptually to form classes included in other classes, and he can relate numbers mathematically, by adding or subtracting them. This is to say that he now has the mechanism which will release him from the directly perceived world of objects, and actions upon objects. Instead, he can now operate upon symbols which stand in place of the environment, these symbols being public rather than private.

Throughout the period of concrete operations, the child extends and perfects his operational activity over a wide area. Looking back to the modes of thinking common to the pre-operational period, which were the preconcept, transduction, juxtaposition, syncretism, centration, and static representation, it can be seen that it was their nature to be inflexible and rigid. Their metamorphosis to mental operations, occurred by a gradual loosening and by increased mobility, until finally they became reversible.

SUMMARY

Around the years of seven or eight, the processes of concrete operational thinking begin to emerge. These operations are mental actions, derived in the first place from physical actions, which have become internal to the mind. By virtue of concrete operations, 'immediately given data' can be restructured in the mind into new forms. Contact with the environment is maintained during such mental actions, because by reversing them a return to the perceived form is always possible. Concrete operations are reversible in two ways, by the inversion of combinations (classes), and by the reciprocity of differences (relations.) Reversibility permits conservation.

RELEVANT SOURCES

1. *The Child's Conception of Number.*
2. *The Early Growth of Logic in the Child.*
3. *The Child's Conception of Space.*
4. *The Child's Conception of Geometry.*

The progress of concrete operations: the threshold of formal operations: from about nine to twelve years

The emergence of concrete operations has been examined in the previous sub-section. During the second half of this period, a number of developments can be seen which eventually lead to the appearance of formal operations. One such development has already been mentioned—the successive achievements of the various conservations. Other changes will now be outlined.

COMPLEX CLASSES AND LINKED STATEMENTS

This development is illustrated by Piaget in an experiment with flexible rods. The children are given a bath of water and a number of rods which differ in material, length, thickness, and cross-sectional form. The rods can be pushed into one side of the bath, through holes, so that they extend over the water. Three different weights are provided which can be placed on the ends of the rods, as shown below. The children are questioned about the factors which affect the flexibility of the rods. These factors can be discerned by adding the weights and noting how the rods bend towards the water surface.

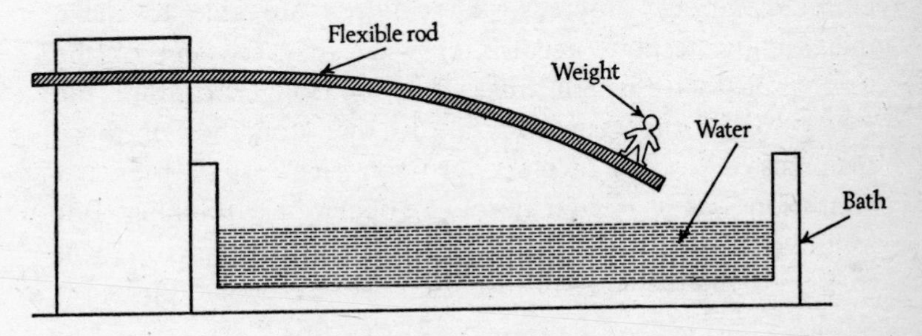

The kind of answers given to Piaget's questions are as follows:

At 9:2 (Child) 'Some of them bend more than others because they are lighter (he points out the thinnest) and the others are heavier.'

(Adult) 'Show me that a light one can bend more than a heavy one.' He was given a short thick rod, a long thin one, and a short thin one. He took the long thin rod and the short thick rod and placed the same sized weight on both. He was asked why he chose those two rods from the three, and the reason he gave was that, 'They are more different.'

At 10 : 9 (Adult, to another child) 'Can you tell me without trying whether that (weight) will reach the water with this rod?'
'It could, but by pulling it in a little' (i.e. by keeping the rod long over the water).
'It's made of the same metal as the other one but it is thicker, so you wouldn't have to pull it in as much as the other.'
(*Logical Thinking*, p. 47 et seq.)

Piaget's analysis of these answers, and many others not quoted, can be summarized as follows. As the period of concrete operations proceeds, the children are able to make increasingly accurate records of what they see and of the results of their experiments. They no longer confuse the behaviour of the materials, due to the properties of those materials, with their own actions upon such materials. Such a confusion was characteristic of pre-operational thinking. For example, a younger child might say that a brass rod bent easily, although this result occurred because he had pushed it harder.

The child's ability to handle classes and relations becomes skilled. He can form complex classes like long thin rods, thick square brass rods, etc. He can make statements which link such classes together, such as 'the long thin rod bends more than the short thick one'. He becomes aware that many factors may be at work in a given situation, and he will attempt to separate one from another. In the flexible rods experiment, for example, reversibility by inversion allows him to either try a brass rod or not a brass rod—say a steel one—or to add a weight or remove a weight. But this method of eliminating factors cannot remove the effect of length or cross-sectional form, and thus he does not completely succeed. The ability to form complex classes, and to link them in statements, gives rise to a situation which runs counter to the child's purpose.

The wealth of information he can now elicit from the situation produces its own confusion, and the child resorts to haphazard experiment, hoping that something will turn up.

THE REVERSIBILITY OF THOUGHT

This development is illustrated in Piaget's experiment with the hydraulic press. The children are given a simple hydraulic press. They have four different weights in boxes, of the same appearance, which can be placed on the piston head. There are three liquids, of different density, which can be placed in the hydraulic system. The children are questioned about the forces at work in the system, and their relationships to each other. These may be classified as follows:

Action Forces:

1. The effect produced by adding weights to A.
2. The effect produced by removing weights from A. (The inversion of 1.)

Reaction Forces:

3. The effect produced by the displaced liquid in B, due to its height and density. (The reciprocal of 1.)
4. The effect produced by substituting a less dense liquid. (The inverse of (3) and the reciprocal of 2.)

It can be noted that the inverse of (4) is (3), and the reciprocal of (3) is (1). Thus (1) is the reciprocal of the inverse of (4), or the correlative.

The kind of answers given to Piaget's questions by children at the concrete operational stage are as follows:

At 10: 3 Before adding a weight, the child says, 'It's going to rise here (B), and here (A) it will go down.'
The child then performs the experiment, and remarks, 'Oh! I thought it would go up higher.'
(Adult) 'Why?'
'Because the piston didn't go all the way (down) and the water didn't go all the way up.'

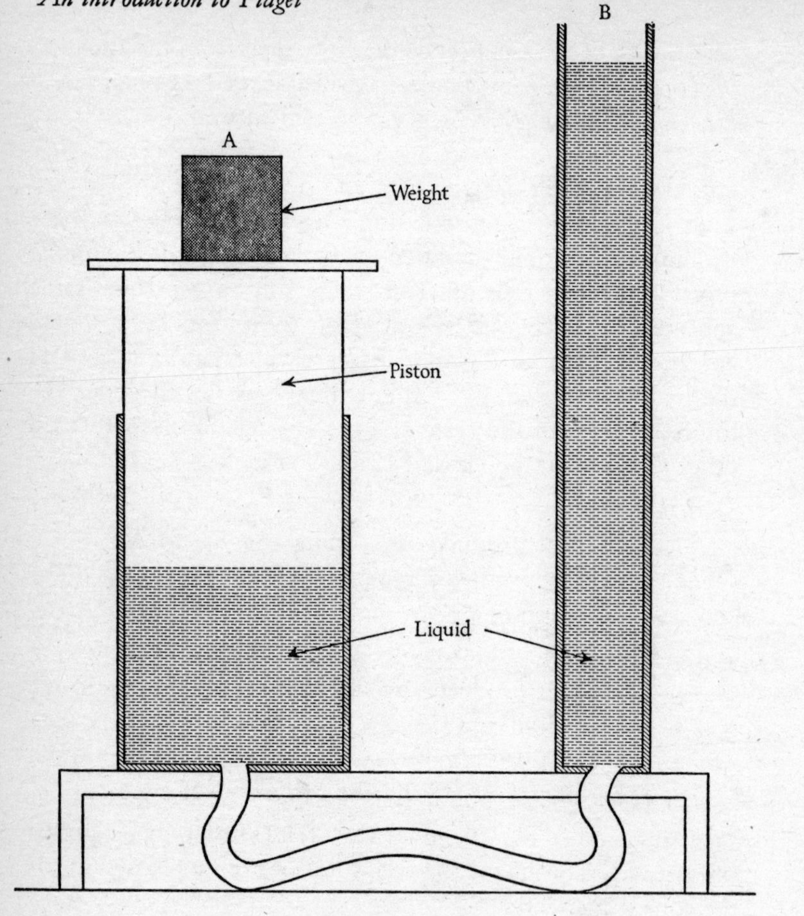

(Adult) 'And with that box?' (A heavier weight.)
'Way up there.' (Higher in B.)

At 10: 10 (Another child, before the weight is added) 'The tube (piston) is going to fall and the water will overflow . . .'

At 11: 0 (Another child experiments first with the alcohol, and then with water.) He concludes, '. . . When the liquid is heavy, it has more weight, more pressure; it goes down faster here (in A).'

(Adult) 'And here (in B) the liquid doesn't press?'
'No, since it's this one (in A) that goes there (in B).'
(*Logical Thinking,* p. 149 et seq.)

As expected, the children can note accurately what they see, and can make linked statements about the results of their experiments, yet they are unable, at present, to understand how the hydraulic press works. Piaget's analysis of the reasons for this can be summarized as follows.

The equilibrium reached in a hydraulic press depends upon the balance of forces acting in the system, i.e. the equivalence of action and reaction. The effect of adding or removing weights at A is always equated with the effect of the weight of the displaced liquid in B. To understand the working of the system, it is necessary to understand both the effect of adding or removing weights (a thought process requiring reversibility by inversion), and the effect of the reaction forces caused by the displacement of the liquid (a thought process requiring reversibility by reciprocity.) It can be seen from the answers quoted above that the children expect the liquid in B to go higher, or to overflow, and that the piston at A will go down further or faster. These answers show evidence of inversion but not of reciprocity. An understanding of the hydraulic press requires thought processes which display both inversion and reciprocal reversibility at the same time. This integration does not appear in the period of concrete operations.

SUMMARY

Developed concrete operations permit accurate observation and experiment. Complex classes can be formed and linked statements can be made. Two important limitations are inherent in concrete operations. No system has yet emerged by which the child can interrelate his classifications, and the two forms of reversibility are not yet able to work side by side.

RELEVANT SOURCE

1. *The Growth of Logical Thinking.*

The emergence and development of formal operations: from eleven years into adolescence

The nature of the changes which occur, as formal operations emerge from concrete operations, can be illustrated by reference to the two experiments outlined in the previous sub-section.

THE EMERGENCE OF SYSTEMATIC THINKING: THE FLEXIBLE RODS EXPERIMENT

At 11: 10 (Adult) 'Could you show me that a thin rod bends more than a wide one?'

The child takes two rods of the same metal, the same cross-sectional form, and the same length, but of different cross-sectional size. He puts a 100 grams weight on the thick rod, and a 200 grams weight on the thin rod. He concludes that the thin rod bends more.

(Adult) 'Is that way right?' (He has not kept the weight constant in both cases.)

The child changes one of the weights so that they are now both the same. He then replies,

'You see, this is the right way.'

At 16: 10 (Adult to another child, after experimental trials)

'Tell me first what factors are at work here?'

'Weight, material, the length of the rod, perhaps the form.'

'Can you prove your hypotheses?'

The child takes a steel rod and compares the effect of a 200 grams weight and a 300 grams weight on the rod. She then says, 'You see, the role of weight is demonstrated. For the material, I don't know. I think I have to take two rods of the same form. Then to demonstrate the role of the metal I compare these two . . .' (She selects two rods in which all the variables are the same, except the metal.) (*Logical Thinking*, p. 57 et seq.)

Important changes in the thought processes are apparent in these answers. It is clear that the child is no longer overwhelmed by the information which his observations and experiments produce. He has created a method for dealing with it. It is this method for processing information which gives a clue to the underlying nature of formal operations. It can be seen, in both examples, that the child can now not only separate the variables, but he has also discovered a means by which one variable can be examined at a time. This is achieved by holding all variables constant except the one to be observed. The younger of the two children quoted above can only just achieve this. However, both these children display an air of organization in the way in which they approach the problem. They spend some time trying various rods, they form a tentative hypothesis about the factors involved, and they then verify or refute their theory. Their thoughts about the problem are expressed in statements which have a propositional form; 'If I do x and y results, then x has caused y, other things being equal.' The child's thoughts are no longer tied to the task in hand, in that his mental actions are not dominated by the reality before him. On the contrary, the reality of the situation is one of a number of possible situations which his thought processes can conceive.

THE INTEGRATION OF INVERSION AND RECIPROCITY: THE HYDRAULIC PRESS EXPERIMENT

At 11:6 (Adult) 'Why doesn't the piston go all the way to the bottom?'

'Because the piston no longer has enough force to bear down. It is held back because the liquid is heavier than the piston.'

At 14:6 (Another child) '... the water is dislodged by the weight (in A): It comes into equilibrium at a certain moment because the weight of the water (in B) increases when it rises.'
'What makes the water rise?'
'The weight of the boxes; it makes a greater pressure at the bottom of the tube and that dislodges the water.'
(*Logical Thinking*, p. 159 et seq.)

It can be seen that the children are now aware of the reaction produced by the weight of the liquid in B, and of the equality of action and reaction. This understanding involves the use of thought processes which employ both inversion and reciprocity at the same time. This integration of reversible forms of thought is a basic property of formal operations.

THE NATURE OF FORMAL OPERATIONS

Piaget's analysis of formal operations achieves a number of things. It explains how formal operations grow out of concrete operations, it describes the structure of such operations, and the consequences which will follow in terms of intellectual achievement, because of those structures. This analysis can be outlined as follows.

The child's increased ability to form complex classes out of the properties of things, and to make linked statements about them, produces, in the end, a wealth of information which he is unable to understand. Piaget remarks, 'Sooner or later they have to retrace their steps, for if too complicated linkages are built up, the variables left unanalysed at one moment will later reappear as disturbing influences.' (*Logical Thinking*, p. 286.)

When this situation arises, the child attempts to rearrange his information in order to simplify it. He discovers that this can be achieved by keeping some variables constant while experimenting with the others. He neutralizes the effect of all the factors, except the one he wishes to study. This approach is a manifestation of reversibility of thought by reciprocity, for to equalize the differences is to produce an equivalence. Thus, with the emergence of what Piaget calls 'the schema of all other things being equal', the reciprocal reversibility of mental actions occurs together with reversibility by inversion.

This new approach produces new kinds of class relationships. These relationships are not basically different from those which are formed by concrete operations. They differ only in the following way. Instead of combining the properties of objects into classes, they combine the classes, thus formed, into other classes. In effect, the child now combines classes together in a way similar to the method he previously used to combine the properties of objects together. For example, a concrete operation will produce a statement like, 'The rods which are long and thin, bend more than the rods which are short and thick.' On the other hand, a formal operation will produce a statement like, 'If the long rods bend more than the short rods, other things being equal, then greater length causes more bend.'

This step forward is important because in the process of combining classes, the child gradually produces a complete system of all possible combinations. This complete system begins to work independently of its content, and becomes an autonomous thinking instrument which the child can apply to data of the most diverse kind. Piaget refers to this new organization of mental actions as the combinatorial system. It has the vital property of being a self-regulating and sustaining totality. Formal operational behaviour is the product of this combinatorial system. Formal operations produce changes in the child's attitude to the environment. He now has a powerful problem-solving mechanism at his command. He can use the hypothesis, experiment, deduction approach when investigating his environment. He can treat the particular situation as one reality amongst a range of possibilities which the combinatorial system throws up. He can reason from particular to general and back again. He can make statements having propositional form. He is no longer tied to his environment, for with formal operations he is performing actions, not with the environment directly, but with statements about the environment.

The difference between these operations and those of the concrete operational period is that a concrete operation is a mental action in which classes of objects, or relationships between objects, are combined or related to make statements about the environment, while a formal operation is a mental action in which the statements themselves are combined to produce new statements. The result of this is a further release from the world, for the adolescent is now performing operations on the results of other operations. Such second degree operations express, in another way, a general characteristic of formal thinking, and this can be explained as follows. The child thinking with concrete operations, structures only the reality on which he acts and so extends the real in the direction of the possible. With formal operations, on the other hand, the given environment can be treated as one of a number of possible conditions. The adolescent then verifies which condition actually pertains in the given situation, i.e. he begins with the possible and proceeds towards the real. Thus, formal operations reverse the relationship between the real and the possible. Piaget comments about this as follows:

> The most distinctive property of formal thought is this reversal of direction between reality and possibility; instead of deriving a rudimentary type of theory from the empirical data, as is done in concrete inferences, formal thought begins with a theoretical synthesis implying that certain relations

are necessary and thus proceeds in the opposite direction. (*Logical Thinking*, p. 251.)

As far as the two forms of reversibility are concerned, these mental actions are now combined into a single system which works as a whole. This system is a group of transformations which are, inversion, reciprocity, the inverse of the reciprocal (or the reciprocation of the inverse) and the identity transformation. (The INRC group.) By means of a complex argument, Piaget demonstrates how the group of transformations and the combinatorial system are mutually reinforcing in their properties, the growth of each urging on the development of the other. The result of this is an integrated system in which the combinatorial system and the group of transformations work as one. This organized whole is the structural base from which mature formal operations derive. Piaget suggests that the structured whole begins to show itself at around fourteen to fifteen years of age.

The side-effects of the structured whole, in addition to those already mentioned, permit the adolescent to understand relationships such as those found in mathematical and metrical proportion, probability, correlation, conservations of the second order, and the manipulation of formulae.

SUMMARY

The increased complexity of information about the environment, which developed concrete operations produce, urges a reformation of those structures. The schema of 'all other things being equal' is employed, and classes are combined with classes. A set of all possible combinations emerges, which Piaget refers to as the combinatorial system. The separate forms of reversibility present in concrete operations become integrated into one system. Both these changes are themselves integrated during adolescence to form a structured whole. This total system produces mature formal operations. The by-products of this are: a reversal of thought in which the real becomes a special case of the possible, propositional thinking, and the hypothesis-deductive strategy. Further conservations appear, which require formal operations for their existence, e.g. volume, inertia, etc.

By way of final comment to this Section as a whole, it could be said that Piaget's psychology of intellectual development is a good deal more complex, and considerably richer, than the outline here suggests. It is hoped that the reader will be tempted to turn to Piaget's writing and to sample this richness at first hand.

Section two

Some basic theoretical concepts

This Section deals with some, but not all, of the basic concepts which Piaget has developed to give coherence to his explanation of the nature of intelligence. Some of these have been touched upon already, for example, the view that mental processes are internalized actions and that they are reversible. There are other more general ideas, however, which unite the whole theoretical framework, and it is with these that this Section is concerned.

THE CONCEPT OF INTELLIGENCE AS A DEVELOPMENTAL PROCESS

Piagetian psychology is developmental. This aspect of the hypothesis is of central significance, for it is Piaget's purpose to explain in a self-supporting and logically consistent manner the way in which a newborn baby, totally ignorant of the world into which he has been precipitated, gradually comes to understand that world and to function competently in it.

To clarify the meaning of the term 'developmental' in Piagetian psychology the following analogy can be used. (This analogy may also explain the experimental method which Piaget uses.) Imagine a fictitious biologist who is intrigued by the problem of how a cell divides. The biologist knows that a cell can do this, and he wants to find out how. He therefore takes a cell, puts it under a microscope and observes all he sees. Two difficulties are immediately apparent. First, the parts of the cell are so indistinct that he cannot distinguish one from another, and second, the changes which he can see are so complex that he is unable to record them. To solve the first problem, he adds a number of staining agents to his cell and looks again. The result is better. His second difficulty he solves by photographing the cell several times as it changes. He then examines his photographs carefully, comparing one with another. The pattern of the changes in the cell are now clearer to him. He repeats the experiment with a large number of cells to be sure that they are all behaving the same way. From his pile of photographs of different cells, he picks the clearest pictures and then arranges them in a time sequence which shows the changes that occur.

The result of this study will show an idealized set of pictures, in which the cell arrangement at one moment is the direct outcome of what happened just previously, and a necessary precursor to what happens next. The biologist will explain that his choice of pictures was governed by the fact that the change each of them showed was sufficiently different to be regarded as a new arrangement of cell parts, and that, because of this new arrangement, the cell will behave in a slightly different way. He might go on to say that if he had chosen different staining agents, he might have had a different pattern, but that he is sure his pattern is more or less correct because it makes sense. He would also not rule out the possibility that different temperatures or nutrients for his cells might have made the process of division quicker or slower, or even different.

Now there are a number of similarities between the way the biologist worked on the problem of cell division and the conclusions he reached, and Piaget's hypothesis concerning intellectual development, although it must be clearly stated that it is not being suggested that there is any connection between cell division and intellectual development. Piaget set himself the task of explaining how the mental structures of the new-born baby become the structures of the adolescent intellect. He knew these two extreme conditions were not the same and that there must be changes in between which would explain how the first condition became the last. The problem was to find out what the changes were, how they occurred, and why they occurred.

The first difficulty was to find some staining agent which would highlight the patterns in the mind at various intervals between birth and adolescence. His choice of staining agents was a number of highly original experimental situations into which he placed children of varying ages. Some of these situations were outlined in the previous Section. As the children worked at his experiments, he asked them probing questions to clarify various aspects of their thinking which the experiments produced. From the results of these experiments and observations, he abstracted the fundamental patterns which he discerned must underlie the responses of the children, and these patterns he then pieced together in a way which made sense to him, i.e. that one pattern was the necessary precursor to the next, and so on. The patterns were then arranged in a time sequence, using chronological age to divide one pattern from the next.

Thus we would expect to find in Piaget's description of the stages in intellectual development that they are a logical self-supporting series of changes which can be set out with approximate chronology, allowing a leeway of a year or two here and there. We would also expect that although we cannot pin down the stages too clearly, we cannot, under any circumstances, change the order of progression, for this would make a logical nonsense of the whole sequence. The division of the sequence into patterns, or stages, gives rise to the question of what constitutes a new stage in a process of continuous change. Piaget's answer to this question is complex, because it involves not only the changing organization of mental structures, and therefore overt behaviour, but also the equilibrium state between those structures and the environment. The subject of equilibrium states will be introduced later. As far as the changes in structure are concerned, Piaget suggests that throughout the whole sequence adjustments to those structures are taking place. At certain moments the structures interact with one another and form new patterns. When this occurs, behaviour becomes sufficiently different from earlier behaviour to justify a stage label. Piaget uses the term 'period' to describe a major span of development, and the term 'stage' for smaller spans within a period. This was illustrated in Section 1. The major reorganizations of mental structures occurred with the emergence of sensori-motor thought, concrete operational thought, and formal operational thought.

Although chronological age is used to give the span of the periods, this criteria is approximate and can be treated as a general guide only. The order of stages must be regarded as fixed, but there are ways in which variation can occur. It does not follow that all children will reach the end of the sequence. It has been shown that severely mentally handicapped children may not acquire concrete operational structures, while other children, less handicapped but of below average ability, may not acquire formal operational structures. It can be noted also that, although each stage provides the foundation for the next, this does not mean that earlier modes of thinking disappear. Thus an adolescent can think concretely as well as formally. He may even use pre-conceptual thought at times. A sensori-motor understanding of the world is also still at work. Under certain circumstances, perhaps in the face of a new experience, a child may employ modes of thinking which are developmentally more primitive than those of which he is normally capable. A kind of intellectual regression may occur.

THE CONCEPT OF INTELLIGENCE AS AN ADAPTATION PROCESS

It could be said that Piaget begins his thesis that intelligence is a process of adaptation with a question. If the brain, source of the intellect, is a living part of a living organism, will it then exhibit the properties which are common to all other living organs and if it does, will this fact be a clue to explaining how it functions? Piaget's answer to both parts of this question is 'yes'. The reasoning goes somewhat as follows. The liver is a living organ in a living body, so is the heart and the brain. Although each of these organs has quite different organizations and functions, they undoubtedly have some basic common properties which derive from the fact that they are alive. The same line of reasoning can be extended outside the human body to include all other living things, for the wallflower, the earthworm and the human body are also alike in being alive. What, then, is the essence of this likeness which is exhibited by living things as a whole, including man and his brain? Thus Piaget peers deeply into the functioning of all living things, stripping away all the layers of individuality which organisms, and their parts, express until he comes to the concept which might be called 'livingness'.

The basic principles of this concept which he produces are:

1. There is complete interdependence between a living organism and the environment in which it lives.
2. The organism and its environment are engaged in a continuous process of action on, and reaction to, each other.
3. There will be a balance or equilibrium relationship.

These ideas are not original, but are well known as the concept of biological adaptation. What Piaget does, is to take the concept of biological adaptation and apply it to the development of intelligence in each individual as he matures from infancy to adulthood. Thus the mind functions by employing the principle of adaptation and produces structures which display themselves in intelligent behaviour as a result of countless mental adaptations which have been acquired in the process of ageing. Piaget remarks about this as follows:

> Intelligence is an adaptation. In order to grasp its relation to life in general it is therefore necessary to state precisely the relations that exist between the organism and the environment. Life is a continuous creation of increasingly complex forms and a progressive balancing of these forms with the environment. To say that intelligence is a particular instance of biological adaptation is thus to suppose that it is essentially an organization and that its function is to structure the universe just as the organism structures its immediate environment. (*The Origin of Intelligence,* pp. 3–4.)

Thus intellectual organization becomes a special case of the general life process of adaptation. It can be seen that it is a special case in the following way. Biological adaptation of the organism to the environment requires a constant adjacency of the organism to its material surroundings for the interrelationship of the two to occur. Mental adaptation, on the other hand, permits a progressive release from this adjacency until, in the end, the intellect can function on its own. Piaget remarks: 'It is in this sense that intelligence, whose logical operations constitute a mobile and at the same time permanent equilibrium between the universe and thought, is an extension and a perfection of all adaptive processes.' (*Intelligence,* p. 9.)

Thus the limit condition for the evolution of mental adaptation is the complete structuring of the intellect to encompass the entire universe, actual or potential, regardless of spatial or temporal distance. It will be helpful, again, to look at this idea of intelligence as an adaptation by means of an analogy. Take the simple one-celled animal called the amoeba, which can be found living in the mud at the bottom of a pond, and which is about the size of a pin-head. Like all living things, the amoeba needs to feed in order to live and grow. To obtain its food it moves around in its watery environment in search of nutritious particles. Movement is achieved by changing the shape of its single cell, protruding first one part, then another, in the direction it 'wishes' to go. When it comes into contact with a suitable particle of food, the extended parts of the cell surround the particle, join up, and engulf the food into the body of the cell. The process of digestion now takes place. The amoeba secretes simple digestive juices, which envelop the food particle, and break down the complex compounds in the food and transform them into new compounds. These are then built into the substance of the amoebic cell. Those parts of the food particle which cannot be transformed in this way remain unaffected and the amoeba discards them by moving on and leaving the rejected portions behind.

If this process is examined in more detail, it can be seen that the amoeba takes in the food particle and changes it into new substances which are of a kind that can be used to construct the living matter of the cell. Thus the food must be changed in order that it may fit into the existing structure. The kind of substances which the amoeba can accept will depend upon the nature of its cell and the nature of its digestive juices. This process of exploring the environment, taking in parts of it, acting on them and transforming them into new forms which will fit into its cell, might be called assimilation. What is assimilated, or rejected, will depend upon the nature of the assimilating structure and its requirements at any given time.

Next, it can be seen that the shape of the food particle will effect the shape of the space which the amoeba must make in order to take in that particle, and also that the chemical nature of the food particle will determine the type of digestive juices the amoeba employs to transform the particle into its own living material. Thus the final form of the amoeba and its digestive juices will depend upon the chemical nature of the food available. This process, whereby the condition which surrounds the amoeba dictates its form and structure, might be called accommodation.

These two processes of assimilation and accommodation interact continuously, one with another, and the balance, or equilibrium between them at any given time can be expressed as the adaptation of the amoeba to its environment. Thus the two processes are complementary and inseparable, expressing themselves as adaptation. The form of the amoeba dictates the type of food digested, and the type of food digested dictates the form of the amoeba.

Piaget sees the working of the intellect in a similar way. Every experience we have, whether as infant, child or adult, is taken into the mind and made to fit into the experiences which already exist there. The new experience will need to be changed in some degree in order for it to fit in. Some experiences cannot be taken in because they do not fit. These are rejected. Thus the intellect assimilates new experiences into itself by transforming them to fit the structure which has been built up. This process of acting on the environment in order to build up a model of it in the mind, Piaget calls assimilation.

'Intelligence is assimilation to the extent that it incorporates all the given data of experience within its framework.' (*The Origin of Intelligence,* p. 6.)

It can be seen also that the nature of the environment in which the intellect is working will affect the kind of structures built up in the mind. For the assimilation process can only work upon the experiences available. Now with each new experience, the structures which have already been built up will need to modify themselves to accept that new experience, for, as each new experience is fitted in to the old, the structures will be slightly changed. This process by which the intellect continually adjusts its model of the world to fit in each new acquisition, Piaget calls accommodation.

'There can be no doubt either that mental life is also accommodation to the environment. Assimilation can never be pure because by incorporating new elements into its earlier schemata the intelligence constantly modifies the latter in order to adjust them to new elements.' (*The Origin of Intelligence*, pp. 6–7.)

These two processes working together produce the adaptation of the intellect to the environment, at any given time in the developmental process.

Before leaving the analogy of the amoeba, it is important to stress that there is a fundamental weakness in the comparison. The analogy serves in so far as the taking in of nutrients and the rebuilding of them into amoebic substance is comparable to the intellect absorbing experiences and restructuring them to fit into a mental model (assimilation). Also, that the final form of the amoebic cell depends upon the nature of those nutrients is comparable to the effect which new experiences have in re-aligning the mental model (accommodation). In both cases, the result of the two interacting processes can be called adaptive. However, after this, the analogy breaks down, for it can be seen that the nutrient balance between the amoeba and its environment does not change qualitatively during the life of the creature (i.e. it is not developmental), whereas in the case of mental structures, there are changes as the individual grows older. As each new experience finds its place in the mind, and modifies the old experiences, the intellect becomes slightly more comprehensive, and the mental model of the external world becomes more complete.

The growth of the intellect is an accumulative process but new experience is not stuck on willy-nilly, for it amalgamates with what is already there, thus changing it and being itself changed. There are 'occasions' during this process when the mental structure seems to realign itself, or shake down into a more viable system; stage changes. At any given time in this developmental process, there will be an adaptive equilibrium, which becomes more and more broadly based with succeeding stages. In discussing the theory of assimilation, Piaget writes:

will have built up sensori-motor representations of these objects in the mind, by virtue of the actions he has performed with them.

Now a new toy is introduced, a red cloth ball. He will immediately begin to adapt to the properties of the red cloth ball by playing with it in all the ways described above. He will fit this new object into the mental model of the ball he has already created. It has the same colour and shape as the other ball, and it rolls. Thus he assimilates new into old. At the same time, he will discover that it does not bounce in the same way as the red rubber ball, that it is soft and warm like the teddy, and that it does not make the same noise when dropped. He will therefore need to adjust his mental model of the ball to incorporate the cloth one. He will accommodate his idea of the ball by modifying it to include softness like teddy, but shape and colour like the rubber ball. Thus he accommodates old to new.

His adaptation to the idea of a ball is now wider and more stable. What will happen when a red balloon is introduced? It is the same shape and colour as the other balls, and it will roll and bounce, but it is very squeezy and gives a different motor response (weight). All these properties could be assimilated into existing structures, and the accommodation process would not be difficult. However, he would then discover that when he pushed it away it moved upwards and not downwards, like every other object he had played with. This behaviour would require very considerable accommodation if it were to be fitted in. The model would have to be readjusted to allow for things to go upwards as well as downwards. He might experiment with his other toys, and find that under certain kinds of push, they would also go upwards before falling. The adaptation could be acquired after a considerable accommodation process.

Imagine now that the balloon bursts. There is a loud noise and the balloon has gone! This experience is quite unique to him, and he rejects it, for there are no assimilatory structures into which this experience can be fitted. The disappearance of objects, in the past, has not been accompanied by a loud, sharp noise, and an intervening object has been present. No accommodation can be made to this event because it is too extraordinary and would require a complete distortion of all the structures for it to be fitted in.

Finally, it can be seen that there will be certain properties of the ball to which the imaginary two-year-old will not be able to adapt. He will not adapt to the idea that a ball is a member of a class of three dimensional objects, or that it has a minimum area in contact with the ground, or that it has the maximum volume for surface area. Accommodation to such concepts will occur much later in his mental development, as was outlined in the first Section.

THE CONCEPT OF MENTAL STRUCTURES AND EQUILIBRIUM

Piaget explains intellectual activity by reference to such terms as organization, operational systems, structured wholes, equilibrium, and so on. These terms are difficult to understand and before plunging into Piaget's theory about mental structures, it might be helpful to look at some possible meanings of the terms involved.

The word structure implies organization or discernible pattern, in which there are parts forming a whole. The way the parts are put together can define the structure. The structure of a brick wall could be defined in this way. This would be an example of a static structure. In addition, there can be a static structure in which dynamic occurrences take place. Such dynamic occurrences would themselves have structure which could be described by rules of operation. The way the operations work together would define the structure or system.

An example may make this clearer. A household central heating system has a static structure of organized parts, like the water tank, pipes and radiators. In addition, it has a dynamic system which is expressed in the movement of the water according to the law of convection, which is its rule of operation.

A further property of structures is their stability, or lack of it. Static structures show their stability by their rigidity. Dynamic structures also show stability, but in a different way. In the case of the water in the central heating system, the movement of the water in one part of the system is always compensated for by movements in the other parts. The system as a whole remains stable because of these compensated movements. The important point which emerges from this is that the stability of a dynamic structure derives from the fact that all the changes of which it is capable are in balance, or equilibrated.

To return to the working of intelligence, the structures with which Piaget describes mental development are dynamic, and they are defined by their rules of operation. When reference is made to concrete or formal structures, these are dynamic systems. As for the relationship of such structures to the neural network, Piaget remarks: 'If neurological considerations come to round out our explanation at some later date, the structures of groupings, lattices, and groups will reappear in this new perspective, and, as a result, these laws of equilibrium will prove to be more general than when linked to behaviour patterns alone.' (*Logical Thinking,* p. 333.)

To summarize, in Piaget's psychology of intelligence, mental structures are dynamic and are defined by means of operational rules which, taken together, form an equilibrated system. Although setting aside, for the moment, the question of equilibrium, it is an important aspect of the theory that some equilibrium conditions are more stable than others. There are two other points to note about these structures. Firstly, they change during the intellectual ontogeny, and, as a consequence, the form of the equilibrium changes. Secondly, the total structure has sub-structures which themselves show operational properties.

SCHEMAS

Reference to sub-structures leads directly to a consideration of what Piaget calls a schema (plural: schemas or schemata). The schema is an ever-present term in Piaget's theory of mental structures. It appears throughout the developmental sequence, at all stages. Its precise nature is difficult to pin down, but it does not violate the theory if one thinks of it as a substructure. Thus, within the total structure there are schemas and the behaviour of the structure as a whole, changing as it does from period to period, results from the way the schemas work and the way they are organized amongst themselves. There are, therefore, two aspects of schema behaviour which require explanation; first, how a schema itself works, and second, how schemas are organized amongst themselves.

Sensori-motor schemas have their origin in the reflex behaviours with which the baby is born, such as the tendency to suck, grasp, or cry. The basic property of such reflexes is repetition. This property requires environmental stimulation to set it going, but once it has begun, the reflex schema will exercise and strengthen itself on any suitable object. Piaget refers to this property of repetition as 'reproductive assimilation'. In connection with the sucking reflex, Piaget writes: 'Thus, according to chance contacts, the child, from the first two weeks of life, sucks his fingers, the fingers extended to him, his pillow, quilt, bedclothes, etc.; consequently he assimilates these objects to the activity of the reflex.' (*The Origin of Intelligence*, p. 34.)

This example can be used to illustrate two other basic properties of schemas. By repetition, the schema incorporates varied objects into itself, and thus its field of application becomes wider. Piaget calls this property 'generalizing assimilation.' However, because of the varied qualities of the objects assimilated, a converse action takes place. This action is schema differentiation, in which the schema responds differently to the various objects it assimilates. Piaget's term for this is 'recognitory assimilation'. This differentiation produces motor recognition.

To summarize, the basic activity of a schema is (1) repetition, and given this, the schema, (2) generalizes, because a variety of objects are capable of satisfying the repetitive process, and (3) differentiates, as a result of that variety. This total process produces an organized whole, or a sub-structure.

The changes described above can be thought of as occurring simultaneously in many schemas, each of which has a reflex tendency at its base. At certain times in the development of schemas, new behaviours can be observed in the baby. Piaget describes one of these as follows:

> Now comes the moment when the child looks at his hand which is moving. On the one hand, he is led, by visual interest, to make this spectacle last—that is to say, not to take his eyes off his hand; on the other hand, he is led, by kinesthetic and motor interest, to make this manual activity last. It is then that the co-ordination of two schemata operates, not by association, but by reciprocal assimilation. (*The Origin of Intelligence*, p. 107.)

This example describes how schemas become organized amongst themselves by a process of assimilating each other. Piaget calls this 'reciprocal assimilation.' The result of the two schemas assimilating each other is a new mode of activity in the environment; in this example, the co-ordination of hand and eye. In terms of mental structures, reciprocal assimilation produces a new totality, or a more elaborate sub-structure. The process of reciprocal assimilation is of vital significance to Piaget's theory of intellectual development, for this process is used to explain the formation of complex co-ordinations between schemas. When schemas assimilate each other, they create a new intellectual action, a new way of thinking, a new way of representing the relationships found in the environment, and thus a new way of behaving with the environment. When the baby exhibits a behaviour pattern hitherto absent, or solves a problem which earlier had been beyond him, the 'inventive act of intelligence' which has occurred is explained as the result of the reciprocal assimilation of schemas. 'In short, invention through sensori-motor deduction is nothing other than a spontaneous reorganization of earlier schemata which are accommodated by themselves to the new situation, through reciprocal assimilation.' (*The Origin of Intelligence,* pp. 347–8.)

The discussion so far has been based upon Piaget's comments about schema behaviour in the sensori-motor period. However, the term schema appears throughout his writing, and the combination of schemas, or their re-alignment into new totalities which work as a more or less permanent system, can be found in the Intuitive stage, or in the Concrete and Formal Operational periods. Piaget's explanation of the progress of intellectual development often contains reference to the emergence of new schema behaviour derived from existent schemas. For example, in the late Pre-operational period, when the child begins to grasp the notion that two different spatial arrangements are composed of the same number of counters, this intuitive understanding is explained as a result of the emergence of a new schema.

'The child's success . . . is achieved through a new semi-operational method, or rather through the development of a schema which already existed within the global comparison, but which comes to the fore at this level . . .' (*Number,* p. 88.)

The period of formal operations is marked by the presence of several new 'operational schemata,' such as those connected with proportion, correlation or mechanical equilibrium. The presence of the schema, 'all other things being equal,' is regarded as a hallmark of formal thinking. These operational schemata are seen as the result of the integration of the many concrete operational groupings into the combinatorial whole of formal thinking. Thus, throughout Piaget's writings, the combination of mental sub-structures among themselves produces new modes of thinking. In the first years of life, the reciprocal assimilation of schemas produces new modes of behaviour. This is continued in later years and structural reorganization occurs to produce powerful problem solving strategies. In summary, schemas might be described as substructures, dynamically organized, which show the properties of reproductive, generalizing, recognitory, and reciprocal assimilation. The activity of intelligence, and its ontogeny, result from the constructive processes of schemas worked out in contact with the external environment, and in accord with internal schema organization. Piaget comments as follows: '. . . schemas, being instruments for adaptation to ever varying situations, are systems of relationships susceptible of progressive abstraction and generalization.' (*Play, Dreams and Imitation,* p. 99.)

STRUCTURAL CHANGE AND EQUILIBRIUM

The basic property of schemas is assimilation. They assimilate the environment and each other. Piaget stresses this by referring to assimilation as the basic fact of psychic life. Given this property, mental structures tend to become increasingly elaborate throughout the sequence of periods. Within each period, the structure gradually builds up by a multiplication of schema relationships until there occurs a re-grouping of relationships and the structure exhibits new forms of organization. This re-grouping then forms the basis for further assimilation. Thus, the sensori-motor period reaches fruition when a system of motor actions permits the conservation of objects and the formation of the group of displacements. From that time onwards new assimilations fit themselves into the schemas of conserved objects (symbolic schemas), and then structural multiplication continues afresh. Pre-operational fruition is reached when symbolic schemas group themselves, and the concrete conservations appear. The groupings of classes and relations then proliferate until they become reorganized. This produces the combinatorial system and the transformation group of formal operations.

The mental structures of each period have their own equilibrium condition, each more stable than the last. Piaget remarks that, 'Only intelligence, capable of all its detours and reversals by action and by thought, tends towards an all-embracing equilibrium by aiming at the assimilation of the whole of reality . . .' (*Intelligence,* p. 9.)

This equilibrium is derived, in the main, from two factors—the reversibility of the operations performed and the content to which they are applied. Sensori-motor equilibrium shows reversibility of motor actions, and the contents are real entities. Concrete operational equilibrium shows reversibility of classes by inversion, and of relations by reciprocity. The contents are the concrete properties of the environment. Formal operational equilibrium shows the integrated reversibility of the group of transformations, and the contents cover the possible as well as the real environment.

A further point can be noted about equilibrium growth. The equilibrium of mental structures can be taken to mean an equilibrated system of relationships between mental actions and environmental occurrences. These relationships are achieved through the assimilation and accommodation process. There are times during every period when the equilibrium of the system as a whole, regardless of its relative stability compared with other periods, is temporarily disturbed. It is, in fact, necessary for structural growth that these disturbances occur, for in restoring the equilibrium the stability may be increased. The loss of equilibrium occurs when the processes of assimilation and accommodation are not in balance in each period. When assimilation is dominant, the environment is subjugated to the dictates of the mind. When accommodation is dominant, the reverse occurs. Piaget has traced this inbalance in each period.

In the sensori-motor period, primacy of assimilation occurs, for example, when the child sucks his thumb. Primacy of accommodation is indicated, for example, when the baby copies the movements of his mother's hands. In the pre-operational period, primacy of assimilation is seen in symbolic play, when, for example, the child uses a stick to represent a gun. Primacy of accommodation is seen in imitative play, when, for example, the child copies his mother's behaviour. In the concrete operational period, fantasy games indicate primacy of assimilation, and model-making indicates primacy of accommodation. In the formal operational period, fanciful theorizing would show primacy of assimilation, and scientific experiment would show primacy of accommodation.

Adapted intelligence is an equilibrium of the two processes with neither dominant, for example, when a scientist makes a correct deduction after an experiment, or when a child solves a concrete problem which the environment presents.

THE CONCEPT OF ORGANIZATION

Piaget refers to adaptation as a functional invariant. By this he implies that adaptation is a process which continues throughout the whole sequence of intellectual development. An adaptation acquired in the first year of life, and one acquired in the fifteenth year, will both have been achieved in the same way, i.e. through the twin processes of assimilation and accommodation. On the other hand, for example, mental structures are not an invariant for they change throughout intellectual development.

The concept of organization is similar to adaptation in that it is a functional invariant. This implies that intelligence is always organized at all stages of the developmental sequence. Thus, although the structures change, they nevertheless remain organized structures. The two functional invariants of intelligence, adaptation and organization, are closely connected to one another in the following way. Adaptation is the process by which intelligence relates itself externally to the environment, while organization is the process by which intelligence as a whole is related internally to its parts. Piaget remarks that organization 'is the internal aspect of the functioning of the schemata to which assimilation tends to reduce the external environment'. The concept of organization applies not only to the intellect as a whole, but also to the working of its parts, i.e. the schemas. Schemas are organized wholes, and when they combine with one another by reciprocal assimilation, the result is a new organized whole, or totality. Piaget expresses the relationship between the two functional invariants in the following way:

> The 'accord of thought with things' and the 'accord of thought with itself' express this dual functional invariant of adaptation and organization. These two aspects of thought are indissociable: It is by adapting to things that thought organizes itself and it is by organizing itself that it structures things. (*The Origin of Intelligence,* p. 8.)

THE PSYCHO-LOGIC OF MENTAL STRUCTURES

Piaget has shown that his theory of mental structures, and their equilibrium conditions, can be expressed with economy using the symbolism of mathematics and logic. He refers to this form of expression as psycho-logic. Two mathematical ideas underlie psycho-logic. They are lattice theory, which subsumes the logic of classes, relations, and propositions; and group theory, which is concerned with the rules by which sets can be combined.

Psycho-logic demonstrates how sensori-motor actions, which are 'the motor equivalent of a system of classes and relations', can be viewed as having the properties of a group. It shows also, how these actions become the groupings of classes and relations (semi-lattices). These groupings, in their turn, become the combinatorial system (a lattice), and the group of transformations (INRC group). The lattice and INRC group work as an integrated whole. Piaget does not seek to reduce thought to mathematics or logic. He takes the view that these axiom systems are derived from the perfection and formalization of rational thought. He comments that, 'Logic is the mirror of thought, and not vice versa.' (*Intelligence,* p. 27.)

It would not be appropriate to outline Piaget's psycho-logic here, but it would be useful to show how Piaget links group theory to the concept of mental operations. Group theory deals with the way in which a collection of elements (a set), combine with one another. A set can be composed of elements such as movements in space, objects, symbols, or operations. A group is specified by stating a way of combining two elements of a set together, and by defining the rules which govern the behaviour of such a combination. It can be seen that if a child's physical movements in space, his actions with objects, his internal mental operations with symbolic representations, and his operations on operations, can be pulled together under the theory of groups, the result would be an analytical tool of some power for elucidating mental activity. This is what Piaget postulates in his hypothesis that mental activity shows group-like properties. Although in the rules which govern the combination of elements there are variations, which arise from the way the elements are combined or the kind of elements, in general groups have the following properties:

1. The combination of any two elements results in an element which is in the set, e.g. A combined with B produces C which is a set element. (The Closure Rule.)
2. The order of combining the elements does not effect the result, e.g. the result of combining A with B and then with C, is the same as combining A with the result of combining B and C. (Associative Rule.) It may also be the case that A combined with B is the same as B combined with A. (Commutative Rule.) But this is not always so.
3. The set contains an identity element, which when combined with any element leaves that element unchanged, e.g. A combined with I (the identity element) produces A. (Identity Rule.)
4. Every element has an inverse and the combination of an element with its inverse, produces the identity element. (Inverse Rule.) It may be that an element is its own inverse, but this is not always the case.

Piaget believes that mental operations display these properties. For example, when the baby manipulates objects, the movements he performs form a group of displacements. The Closure and Associative rules are obeyed. There is an identity element (no movement). Every element has an inverse, sometimes itself. Turning an object upside down and then upside down again, for example, is the same as no movement. Mental activity with classes and relations during the period of concrete operations, shows group-like properties. Piaget expresses the Inverse Rule in his concept of reversibility, by inversion or reciprocity. The identity requirement reappears, for example, as an empty set, a null difference, a universal set, an equivalence or a tautology. Mental activity with numbers, which children perform in the period of concrete operations, also shows group-like properties. At the level of formal operations, Piaget shows how the two forms of reversibility themselves express similar properties, in the group of transformations. This group has four operational elements: identity, negation, reciprocal and correlative (the INRC group), and obeys group rules. Piaget shows how the INRC group may come into play in operations dealing with mechanical equilibrium, relative motion and proportion, for example. The experiment on page 51 shows the INRC group in a concrete example.

SOCIAL FACTORS WHICH INFLUENCE STRUCTURE FORMATION

The following social factors are regarded by Piaget as having an effect upon structure formation:

1. The language used in a society.
2. The beliefs and values held by a society.
3. The forms of reasoning which a society accepts as valid.
4. The kind of relationships between members of a society.

It is important to define the perspective which Piaget takes when he considers social influences. His interest is not sociological, and therefore he does not elaborate upon the effect which different cultures, or sub-cultures, may have upon structural development (except in the sense of the historical development of societies and the forms of thought societies display). Piaget's concern is with the general relationship of social factors to structure development, rather than the effect of specific cultural forms. His interest is in the effect such factors will have at different stages in the developmental sequence. It is also important to bear in mind that social influences cannot be viewed separately from the other factors which are at work. In his analysis of structure formation, Piaget distinguishes three principal influences. They are the '. . . maturation of the nervous system, experience acquired in interaction with the physical environment, and the influence of the social milieu'. (*Logical Thinking,* p. 243.)

This view seems to imply that the neural system cannot be

regarded as a formless plastic entity which can be moulded by social pressures and the properties of the physical world. Nor can it be regarded as an entity having a crystal-like property, which merely grows according to the dictates of its own nascent structure. The structure is an equilibrium condition, ready at all times, to respond to its own growth by producing new environmental behaviour, and, in return, to respond to the results which this new behaviour evinces in the social and physical world.

Bearing these points in mind, the social milieu will effect structural development through the assimilation accommodation process, in the same way as does the physical environment. There is a difference, however, in that the social climate in which the child grows is qualitatively different at different ages. In the pre-school years, the child's relationship to other members of his social group is in the role of a subordinate to the subordinating adult. During the school years, his relationship is still subordinated to the adult world, except that he has an equality relationship with his peers. In adolescence, the equality relationship becomes more general as adult roles are assumed. Ontogenetically, the influence of the social milieu is expressed in the following manner. The emergence of the symbolic function marks the time when language begins to affect structural development. The child is then exposed to the rules which govern language construction, together with '. . . an already prepared system of ideas, classifications, relations—in short, an inexhaustible stock of concepts . . .' (*Intelligence,* p. 159.)

During the stage of intuitional representations, the child will be at school, and socially engaged with children of his own age. Piaget sees this group membership with equals in age and status, as a powerful influence in changing the intuitional structures into operational structures. He suggests that being a member of a group encourages co-operative behaviour and provides a concrete model of reciprocal relationships. The child must decentre his viewpoint in order to account for the viewpoints of others. The interchange of ideas takes place using words, and, if he is to communicate, he must accept the meaning of those words in the way in which they are conserved by the group as a whole. The child is urged to verify his thoughts by experimenting with them socially, and so to resolve the contradictions which he discovers in them. All these facets taken together assist the grouping of mental structures into operational systems. Piaget suggests that the properties inherent in a social group are similar to the properties of the operational groupings of mental structures. For example, both display the following: a co-ordination of actions, changing relationships which nevertheless maintain a conserved whole, and a reversibility of actions. Piaget comments on this as follows:

> It is in fact very difficult to understand how the individual would come to group his operations in any precise manner, and consequently to change his intuitive representations into transitive, reversible, identical and associative operations, without interchange of thought . . . The grouping is therefore by its very nature a co-ordination of viewpoints and, in effect, that means a co-ordination between observers, and therefore a form of co-operation between several individuals. (*Intelligence,* p. 164.)

Social factors are also at work at the time when formal operations reach fruition. Formal thought permits the adolescent to examine his own life style and that of the society in which he finds himself; to question and debate the beliefs and values he holds and those he finds around him. Peer group interaction assists this, and the adolescent tests out his thoughts with his equals. These thoughts are often removed from social reality, being flights of idealistic fancy, but this is exactly what would be expected as a product of unaccommodated formal operations. Piaget and Inhelder remark that such thoughts have

'... a sort of Messianic form such that the theories used to represent the world centre on the role of reformer that the adolescent feels himself called upon to play in the future'. (*Logical Thinking*, pp. 343–4.)

However, it is social factors which assist the adolescent back to earth. At about this time his thoughts must turn to choosing an occupation and to joining society as an active and equal member.

As far as moral judgements are concerned, Piaget suggests that there is a roughly parallel development of these, alongside the changes in the child's social group status and the decentring of his viewpoints. During the pre-operational period, the child is egocentric (not decentred in his viewpoints), and a social subordinate. He views rules of behaviour as if they were natural laws handed on to him by his parents. The violation of these laws must bring retribution and the form of punishment which authority decrees is beyond question. The motives behind a misdeed play no part in the course of his justice, the more serious the misdemeanour, the more harsh should be the punishment. Piaget refers to this view of the rules of conduct as the morality of constraint. During the concrete operational period, the child decentres his viewpoint and has equal status amongst his peers. Rules of behaviour gradually become a matter of mutual acceptance. There must be complete equality of treatment under such rules, and no account can be taken of special circumstances when justice is administered. With the onset of formal operations, rules can be constructed as required by the needs of the group, so long as they can be agreed upon. Motives are now taken into account and circumstances may temper the administration of justice. Piaget refers to this view of rules of conduct as the morality of co-operation.

In general terms, Piaget sees these changes in social attitudes, status, and personal decentring, as part and parcel of the whole process of structural development and increased stability of intellectual equilibrium.

RELEVANT SOURCES

1. *The Growth of Logical Thinking.*
2. *Logic and Psychology.*
3. *The Psychology of Intelligence.*
4. *The Moral Judgement of the Child.*

Section three

Learning and teaching from a Piagetian viewpoint

In general terms, Piaget's thesis would seem to have educational application in two main ways. Firstly, there is his view of the way the intellect and the environment interact with one another: the process of adaptation, the influence of physical actions with things, social co-operation, and language. This part of the psychology could be relevant to teaching method and the organization of learning situations. Secondly, there is Piaget's developmental sequence with the modes of thinking and environmental aliments germane to each stage. This part of the psychology could be relevant to lesson content and curriculum organization at different ages.

In one sense, seeing educational practice from a Piagetian point of view will not produce unusual implications. Many experienced or gifted teachers have long applied Piaget's concepts in the class-room without giving conscious consideration to the matter. In such cases Piaget's work lends systematic support to what is intuitively understood. In another sense, however, Piaget's work has greater value. The generality of his concepts, and the panoramic view of intellectual development he offers, can comment upon a diversity of educational matters.

At the outset, however, one important point needs to be borne in mind. Piaget explains how the child learns, but he does not discuss in detail how he may best be taught. Therefore the following comments can only be of a speculative kind.

ASSIMILATION, ACCOMMODATION, LEARNING AND TEACHING

Piaget defines intellectual adaptation as 'an equilibrium between assimilation and accommodation, which amounts to the same as an equilibrium of interaction between subject and object'. (*Intelligence,* p. 8.)

The twin processes of assimilation and accommodation are permanent features of the working of intelligence, that is to say they are present at all stages in intellectual development. Adaptation to the environment occurs only when the two processes are in balance, and at such times the intellect is in equilibrium with its environment. With age, the scope of the adaptation becomes greater. As the intelligence develops, the cognitive processes can encompass greater temporal distance, greater spatial distance, greater penetration beneath the surface of things, and greater understanding of the complexity of cause and effect. This penetration occurs in both the physical and social world. The progress of this development is marked by the appearance of more abstract forms of representation in the mind. Each step forward in intellectual development requires the application of what is already understood to that which is not understood, followed by an act of adjustment in which the known is modified by the unknown. The application of past experience to the present is assimilation. The adjustment of that experience to take account of the present is accommodation. The concordance between these two acts is expressed in adapted intelligence. But it can be seen that each step forward can only occur through a loss of equilibrium, and therefore intellectual development is a process of restoring a disturbed balance between assimilation and accommodation.

Every learning situation involves assimilation. This implies that the child can absorb a new experience only by changing it so that it will fit into his model of the world. At the same time, the presence of this new experience will change his mental model. Therefore, every learning situation also involves accommodation.

From the point of view of assimilation, it can be seen that if an experience is to have any meaning for a child, that is to say, if he can make sense of it, then he must be able to fit that experience into his mental model. In effect, all new experiences must be related to experiences which the child already understands, i.e. all new learning needs to be based upon old learning. An experience is meaningful only to the extent that it can be assimilated. As far as lesson content is concerned, the teacher can never know the precise nature of a child's previous experience, although certain generalities about its content are obvious. At the beginning it will be a world in which the child is at the focus, a world of body actions, a world of the immediate, the here and now. It will be a world composed of the superficial properties of objects and the overt actions of others. To move the child from his focus, and yet maintain his balance within bounds, his points of reference within this world must move with him into the new setting.

As far as the organization of the curriculum as a whole is concerned, the child's assimilation of experience differs substantially from the way bodies of knowledge are collected into subjects. The arrangement of knowledge by subjects is the product of mature formal operations and has little place in the mind of the child. He will assimilate experiences without reference to such formal boundaries. When a learning situation is not amenable to ready assimilation, the result may be a core of undigested information, which has no application except to the situation in which it was experienced and which, therefore, does not serve as a growing point. Also, the rigid division of learning situations into tight subject compartments might produce at a later stage knowledge which is not susceptible to cross-fertilization.

So far, it has been suggested that experiences have meaning to the extent that they can be assimilated, although assimilation does not occur without some accommodation. From the point of view of adaptation and possible development, the accommodation aspect is of considerable importance. In the formal educational situation, adaptation and development are of prime concern. One of the principal aims of the teacher will be to present situations to the child which require him to adapt his past experience. The teacher is concerned with facilitating adaptation and assisting the child along the developmental path. It might be said that the child sees the learning situation from the point of view of his past experience, while the teacher is concerned with the accommodation of that past experience to the present situation. If the teacher takes a long-sighted view of a given teaching situation, not only will the immediate adaptation be important, but also its relationship to future developments; for each learning situation is the base for future learning.

The teacher, in this sense, is the organizer of learning situations in which old experience can be accommodated to new, and these learning situations are forward looking. The teacher's aim will be to encourage the child to apply his knowledge to situations hitherto unknown, and, at the same time, to encourage him to use familiar actions in unfamiliar contexts. To this extent, a learning situation will contain within it something which is unknown, or new, or problematical for the child, which he will feel the need to understand. The achievement of this understanding produces an adaptation. Each adaptation the child makes constitutes a discovery for him, or an insight. However, intellectual development is a gradual process rather than a series of leaps from one insight to the next. It is marked out by minute consolidations and extensions of past experience, with perhaps an occasional flash of insight.

The teaching of number and number operations offers a clear example of this measured foundation laying approach. Many of the activities in the Infant School with objects and structured materials are organized so that the actions of ordering, combining, dissociating and equating, are experienced in a variety of contexts. The extent of this variety, and therefore the extent of possible accommodations, is a measure of the teacher's ingenuity and fertility of mind. From the point of view of adaptation and development, the aim is to assist the child to discover that numerical symbols can be used to stand in place of these objects and that the actions of combining, dissociating, and equating, can be expressed in terms of number operations.

In the sphere of creative work, old experience is re-expressed through the creative medium, as in creative writing, painting, modelling, mime, or drama. Creative work will require an accommodation of old experience to the inherent structure of the medium, and thus an adaptation of old experience. It will provide also an opportunity for old experiences to be recombined amongst themselves in new ways and so avoid their becoming inert. Thus learning situations can assist adaptation to the extent that old experience is accommodated to new.

Piaget's psychology suggests two very general principles which have implications for the educational process as a whole. Firstly, intellectual development is a directed process, one of increasing stability of equilibrium and expansion of intellectual scope. It is the teacher who is aware of this and who is concerned with its progress. Secondly, it is the learner who performs the balancing process which determines the rate of development. (See 'Problems and Difficulties', p. 110.) The school is therefore the place where developmental situations are contrived to the best of the teacher's ability, and is also the place where the child can organize, unconsciously, his own adaptation. It is a place for both structured and unstructured situations. The curriculum and the individual lesson can offer scope for primacy of assimilation in creative work, for example, in art, drama, movement, creative writing, or poetry. Such assimilation is possible with materials which have little inherent form, and in social situations which are free from imposed pattern. All free assimilation involves, also, accommodation. Creative situations impose their own restrictions, materials have their limitations, and social interaction develops its own structure. All expressive activity engenders accommodation, producing copies and models of the physical environment and imitation of the social environment.

MENTAL ACTIONS AND PHYSICAL INTERACTION

Piaget comments about mental operations: 'They are actions, since they are carried out on objects before being performed on symbols.' (*Logic*, p. 8.)

Thus, mental actions are created through the manipulation of objects, and are sustained and developed through a continuation of such contact. The overall importance of this would seem to be that although in the end the child may develop systems of operations which are autonomous, throughout the course of this development, and even at its fruition, mental operations need environmental sustenance. Only in this way can the mental action, and the environmental condition upon which it bears, remain in adapted equilibrium. The way in which the Infant, Junior, or Secondary School pupil manipulates his thoughts differs substantially, but the need for physical manipulation in the environment remains unchanged. Every mental action concerning the physical world can be verified only by physical action upon it. From the standpoint of the school situation, the converse aspect is of more importance, perhaps. This is that physical action performed first in the environment provides the nourishment for mental action. For example, the formal operations of the adolescent are far removed from their sensori-motor origins, but action with objects can still be the springboard for mental actions. The myths surrounding great scientific discoveries lend humorous support to Piaget's thesis. Archimedes leaping from his bath, Newton startled by the falling apple, both derived much benefit from their sensori-motor experience!

All this suggests that learning at any age needs contact with concrete reality. Piaget expresses this as follows: 'The subject must be active, must transform things, and find the structure of his own actions on the objects.' (*Piaget Rediscovered*, p. 4.)

The support which mental actions require from physical actions declines during the course of intellectual development. For the child in the Infant School, the physical manipulation of objects is an essential food, words and numbers provide developmental nourishment. In the Junior School, the diet can be a balanced mixture of symbolic and direct action. While in the Secondary School, physical action provides the hors-d'oeuvre to be followed by symbolic manipulation.

The ubiquity of the concept of the group throughout Piaget's thesis, offers a clue to the kind of overt activity which may, theoretically, nourish mental activity. Learning situations could be structured around the properties of a group. Body actions, actions with objects, with words, with numbers and with statements (mathematical or linguistic), are all group-like, and learning situations could pick out these properties and emphasize them. This would imply accentuation of the following kinds of activity:

1. Finding similarities and differences in body actions, actions with objects, groups of objects, words, numbers, and statements (mathematical or linguistic), thus forming sets.
2. Putting each kind of set together in various ways in order to achieve the same result.
3. Reversing such actions with each kind of set to return to the starting point. Discovering if this is always possible.

4. Counteracting, balancing, compensating and equating two separate sets of these.
5. Transforming and rearranging single sets of these and finding out what has changed and what has not.
6. Finding opposites within and equivalences between sets.

To look at teaching situations in this way, does not in any sense imply that set theory should be taught to children.

MENTAL ACTIONS AND SOCIAL INTERACTION

Piaget comments about mental and social activity as follows: '. . . without interchange of thought and co-operation with others the individual would never come to group his operations into a coherent whole . . .' (*Intelligence*, p. 163.)

The significance of this thesis seems considerable as far as the school situation is concerned. As with the physical environment, the mechanism of social adaptation is that of assimilation and accommodation. The emphasis is again upon the importance of action with people, as it is upon action with objects. For development to occur, there must be individual action upon a human group and a response by the group to the individual action. This implies that the child's passive presence with others will not necessarily assist the co-ordination of his mental actions, any more than watching a group of objects will assist in developing an understanding of the relationship of the objects one to another, and to the observer. Social interaction, in the developmental sense, implies a human group in which each individual contributes to the working of the group as a whole and is individually involved in the shifts and changes of balance which occur within it. Such a system of human relationships replicates the properties of mental operations. It allows the child to become, himself, an action within a co-ordinated system of actions.

In respect to lesson planning, this view places great emphasis upon the value of group activity. It also emphasizes co-operation rather than competition as a basic educational tenet. Under these circumstances other developmental factors can be effective. Within groups of children involved in common activity, there is the need to express points of view, to exchange thoughts, and to discuss ways and means. This leads to the need to verify, or justify, individual ideas, to iron out contradictions, or to adjust attitudes. It leads to the need for words which have agreed meanings and agreed relationships with other words. Piaget remarks: 'As far as intelligence is concerned, co-operation is thus an objectively conducted discussion (out of which arises internalized discussion, i.e. deliberation and reflection) . . .' (*Intelligence,* p. 162.)

The contribution of social interaction to intellectual development is a continuing one, and Piaget underlines its effect at the adolescent stage also. Action groups, discussion groups, and adolescent societies, provide situations in which formal operations can be exercised and adjusted, in the same way that group activities assist operational development in the earlier years.

The educational importance of this would seem to be that the influence which society at large can have upon the adaptation of formal operations is directly proportional to the degree of interaction between the adolescent and the wider community. This stresses the need for learning situations in the upper Secondary School which take the adolescent into society and which bring society into the school. It also points towards the possible developmental value and effectiveness of co-operative action between adolescent and adult within the context of the school society. Personal experience of decision taking, and the debate with which it is accompanied, would assist in the development of understanding in subjects like religious knowledge, history, and social studies. In a less direct sense, perhaps, role playing by the pupils in these subjects would also be effective teaching method.

Piaget's thesis about the effect of adult authority upon the internalization of social attitudes and moral values might have classroom implications at all developmental levels. Although he underlines the fact that rules of behaviour handed down to the child by the adult have '. . . great practical value, for it is in this way that there is formed an elementary sense of duty and the first normative control of which the child is capable'. (*Moral Judgement*, p. 409.) He suggests that such rules must also be discovered by the child himself if they are to have meaning. A condition for this discovery is social activity and co-operation in shared tasks. In such circumstances attitudes will become exposed, then discussed and possibly modified. This view suggests a parallel between accommodation to the dictates of the teacher in respect of facts to be rote learnt (for example a multiplication table), and accommodation to the dictates in respect of rules of behaviour. In both cases the facts or rules may be adhered to without assimilation, and may have limited usefulness. Lack of adherence might be a measure of the lack of meaning which such facts or rules have for the child.

The role of the teacher, in this thesis, would seem to be to develop discussion out of concrete situations in order to find the kind of rules which operate during various kinds of social interaction. This approach would entail an elementary form of practical discourse with the teacher as the wise leader.

MENTAL ACTIONS AND LANGUAGE

Words are symbols which stand in place of the tangible and visible world. They may embody concepts which have no concrete environmental counterparts, and they can encompass other concepts to produce wider concepts. Such symbols can be linked into sentences which describe relationships within the concrete environment and between concepts. This collection of symbols with its syntax is not only the means by which thoughts can be exchanged, it has built into it rules of validation, and is the vehicle for the creation of new thoughts. Thus language and thought are intimately intertwined. The progress of a child's intellectual development can be marked out to some extent by the degree in which he can make use of this fine-spun and intricate system of meanings.

Piaget shows that the extent to which a child can assimilate language, and therefore the extent of its meaning and usefulness to him in his mental activity, depends to some degree upon the mental actions he can perform; that is whether the child thinks with preconcepts, concrete operations, or formal operations. He remarks that: '. . . it goes without saying that the child begins by borrowing from this collection only as much as suits him, remaining disdainfully ignorant of everything that exceeds his mental level'. (*Intelligence*, p. 159.)

Piaget regards language as a contributory factor in the development of mental actions, but not one which by itself can be sufficient. Other factors like neural maturation, and interaction with the physical environment, are also essential for such development. Language is seen as influential because its syntactic structure reinforces the combination of mental actions into operational systems. Language therefore plays an important role in mental growth but not one which can be effective in isolation. This suggests that language will be used most effectively when allied to action with the physical environment. Such activity would be best organized in groups, within which there is freedom to exchange thoughts about the activity. In this way, experience gained directly, can also be translated into symbolic form. The concepts which arise from the activity, and the relationships they make when linked in sentences, will be manipulated individually by each child as he exchanges his thoughts with the others. The teacher's role in this would be to act as a resonator which would throw back the sentences and concepts to the children arranged in new ways and with new words added. The linguistic atmosphere for such activity would always need to be a little richer than required. It seems likely that the birth of the various conservations will be hastened by verbal discussion if the time is ripe. In connection with classifying activity, Piaget remarks that '... the use of classes which have specific names is an aid to differentiating between them and forming the hierarchy'. (*Number*, p. 168.)

Language may also be misused in the learning process. This would occur, for example, when children are required to accommodate to words and word relationships before their mental structures can assimilate them. This might introduce into the child's mind a collection of symbols which he cannot manipulate in accord with their implied relationships, and which therefore he will either discard *pro tem* or distort into relationships which do have meaning for him. In general, this results in the parrot-like repetition of rote-learnt statements which serve as a disguise for lack of understanding. In some subjects, like mathematics or science, for example, where one concept acts as a base upon which others are built, premature teaching may cause the child to lose understanding at an early stage and never regain it.

Progress through Piaget's developmental stages is marked by an increased use of language in mental activity. This suggests that the role of action and language in the learning situation might be reversed at the later stages. This would mean that activity with the environment would be preceded by discussion of the possible approaches and results of such action. After the task is complete, the predictions, the activity, and the results, could then be re-examined. The three developmental periods provide a rough and ready yardstick by which to gauge the kind of words and word relationships which would be most readily assimilated. Within a period, the appropriate language forms can be extended as widely as possible.

DEVELOPMENTAL STAGES AND LESSON PLANNING

In a general way, Piaget's developmental schedule can offer guidance upon the kind of lesson content most suited to children of different ages. This application is limited because the stages are not divided by precise boundaries; there are gradual transitions between them. It is also limited because a child's modes of thinking show variation in different experiential situations. In addition, when large groups of children are to be considered, there will be a spread across stages within the group. In a class of top infants, for example, some children will show preconceptual thinking, others intuitive representations, and some will operate concretely. In a class of top juniors, some children will be working with their first conservations, while others may show incipient formal operations. At the upper secondary level, some children will be operating formally and others concretely.

These limitations upon the use of the stage sequence in an educational context, are considerable, but not so great as to make the application fruitless. Piaget has mapped out a developmental path and has placed signposts upon it. The direction in which they point seems clear, even if the distances are indistinct. In an approximate way the theory can say what kind of learning situation will be within the grasp of a child and what kind will not.

Piaget's experimental situations provide a means by which a child's modes of thinking in specific situations can be gauged. The teacher can look for the presence or absence of these modes by watching and talking to a child as he goes about his day to day tasks. Occasionally, it might be worthwhile to set up such experiments as test situations. Information about a child's mode of thinking which can be gleaned in this way would afford another dimension upon which his intellectual progress could be measured.

Quite apart from such diagnostic use, the experiments provide a means by which the teacher can gain a deeper insight into child thought, and they also afford a practical way into the study of Piaget's psychology. Descriptions of these experiments can be found in:

1. *The Growth of Basic Mathematical and Scientific Concepts in Children*, by K. Lovell, published by the University of London Press. This book contains a wide range of conservation experiments, together with some age norms.
2. *Mathematics in Primary Schools*, Curriculum Bulletin No. 1, H.M.S.O. Appendix 7 contains a description of some Conservation Tests.
3. *The Growth of Logical Thinking*, by Inhelder and Piaget, contains a number of experimental situations.

The concept of educational readiness is implicit in the developmental schedule, and there would seem to be two ways of looking at readiness in a Piagetian sense. (See 'Problems and Difficulties', p. 110.)

Firstly, when a child has reached a given developmental stage, say articulated representation, concrete thinking, or formal thinking, he will be ready to exercise the modes of thinking germane to this stage. There would be a readiness for consolidation and generalization of existent mental actions in a multitude of environmental contexts. Since the modes of thinking with which Piaget is concerned cannot be taught by the telling of them, but can only become firmly established through usage, it would seem that the more widely the child employs them, the better.

Secondly, when a child has reached a transitional stage, say emerging concrete or formal operations, he should be ready to gain some benefit from learning situations which might nourish his nascent forms of thought. There would be a readiness for progressive learning situations. It can be noted, however, that Piaget believes that no one factor alone can bring about structural change. Neural maturation, interaction with the concrete environment, social activity, and language, all play a part. The emergence of new forms of thinking result from these factors working together. The contribution which each factor makes in bringing about structural change, and the form of interplay between factors during the change, is not clear. Therefore the psychology cannot give clear advice on how to teach for intellectual acceleration. It does suggest, however, that all the influential factors which are within the teacher's control should be introduced into the learning situation, and emphasis placed upon direct action with the concrete, group discussion, and the use of language.

The remainder of this Section will outline in general terms some of the ways in which Piaget's theories can suggest or support good educational practice at the different developmental levels.

In the late pre-operational period the child begins to co-ordinate his mental actions into groups which work as a whole. Reversibility of thought shows itself spasmodically, and the first conservations appear in fitful fashion. The child begins to distinguish between his actions and the inherent behaviour of objects. He begins to see sequential arrangements as related transformations. Social co-operation begins. Language is internalized.

In the learning situation, group work will give opportunities for co-operation and discussion about the activity in hand. The physical manipulation of objects will assist mental actions. The range of possible activities is immense but they can be orientated, for example, by sorting objects into groups in various ways, serially arranging objects and reversing such series, matching series, transforming shapes by pouring, folding, cutting up, or modelling, decomposing and rebuilding composite objects, work with equivalences using scales or containers, or patterns. In many cases, these activities can be performed using the children as a group. For example, in sorting activities or in making living patterns by changing positions, or in serially ordered displays of body dimensions.

As far as measurement is concerned, the standard units are not understood at this time. Length is best approached through the body as a measure. Standard measures might be used for comparison. Time is still a personal experience. Various events can be grouped and related to the times of day already identified by the child. Long events and short events can be compared. Standard units of weight and capacity can be included amongst all the other objects used in the various activities. Whenever possible, or appropriate, means of recording activities can be explored. This can be done in pictures or perhaps in simple histogram or pictogram format. The teacher can add the words to complete such records.

One of the most important teaching functions at this time will be to supply the language to go with these activities. Knowledge of the appropriate words to fit or describe the activity will allow further discussion about it and help the child to give expression to his actions and internalize in symbolic form what he is doing. Talking with children about their environment, rather than direct action upon it, is also a valuable activity, especially if it is associated with a 'visit to see.' A useful orientation could be, again, sorting and classifying the environment using words. For example, the children's houses could be discussed to find what they have in common. Are churches and shops houses? What word can we use for them? These discussions can be organized in several ways, for example, taking wholes apart—what kind of clothes is John wearing? Putting parts together—what kind of clothes can we wear? Serial orders can be found—things that are very cold, cold, not warm or cold, warm, hot. Records of such discussions can be made in pictures, or as teacher-prepared wall charts. Words can be added to the pictures and so on.

In the early concrete operational period, the teaching approach will need to be very similar to that employed during the stage of intuitive operations, but the emphasis can be slightly changed. Reversibility of mental actions will have become better established and the child will be able to dissociate his actions from the behaviour of objects. In other words, he will be able to combine, dissociate, and reverse actions in the mind as well as with things. This implies that the symbolic recording of mental actions is now possible in terms of mathematical and verbal statements. For example, three plus two is equal to five, five minus three is equal to two, and all these shapes have four sides and are green. The patterns produced by number addition and subtraction, and by various combinations of numbers which are equivalent, can be examined as patterns in themselves. 'Play' with number combinations will be a valuable consolidation exercise. Nevertheless, alongside this abstracted activity the need remains for the generalization of such operations in all possible concrete areas. Standard units of measurement can be employed, although weight will not be conserved. As far as multiplication and division operations are concerned, these can only grow after number conservation has been firmly established.

The less superficial properties of the environment can now be sorted and related, for example, records can be made of floating and sinking objects, of tastes, of sounds, or of hard substances and soft substances. Correspondences of many kinds are possible, for example, between shapes and the number of their sides, or between temperature and the time of day, or between different clothing and different occupations.

Language development continues as an accelerating ingredient in learning. Every activity will have words appropriate to it and the vocabulary will be built up imperceptibly as precise words are added to experience. Concepts will arise naturally but, from time to time, they can be examined abstractly. A word like 'shoe', for example, can be discussed to find the extent of its exemplars, or words like 'cotton', 'wool', and 'nylon', etc., can be considered together as examples of a concept.

Alongside these activities, which require accommodation, there are the creative activities. These have a vital role to play in that they give scope for the assimilations that have already occurred to become externalized again in different forms and contexts. In this process, new accommodations are achieved and perhaps a more stable adaptation. This suggests the need to find expressive situations which run parallel and are linked with activities requiring accommodation. It implies, in another way, the need to keep the various curriculum divisions closely inter-linked. For example, mathematics can be expressed in pattern-making, history in drama, and geography in model-making.

By about the middle of the concrete operational period, the child will have conserved weight and be able to measure area. It is likely that he will have separated time taken from distance moved, and he will be gaining an understanding of perspective. These developments will increase considerably the kind of environmental investigations which are meaningful to him, and which may help him clarify these notions. Model construction of all kinds, three dimensional maps, compass and map work, building up shapes by folding and cutting, and examination of internal sections of solid objects, will all be useful experiences. Work with clocks and rulers can be expanded in many ways. Units of time measurement, from a second to a year or so, can be used meaningfully. Work with time-tables, shadows, primitive clocks, and the pendulum will be useful.

At about this time, children will be able to manipulate multiple classes and relations. This ability permits many kinds of activity. For example, an environmental study of different kinds of vehicles passing along a busy road could be expressed as follows: Large vehicles—articulated lorries, six-wheeler lorries, tipper lorries. Medium-sized vehicles—delivery vans, with or without windows; five cwt. trucks, etc. Correspondences between these classes and their usage, or their speeds or loads carried, could be made. Pond life could be studied by types of movement or by feeding habits, etc. Flora could be studied by habitat. These underlying mental actions could be exercised widely in most curriculum subjects. Concept development would then be spread horizontally and the appropriate language forms and vocabulary would be built up. Group work and source books will reinforce this. When separate groups of children are working in related areas, which is usually the case, an integrated display of some kind can show fresh relationships in concrete form, and afford a basis for class discussion.

Towards the end of the concrete operational period, the child will be able to classify, order, and relate his environment in many complex ways. He will be aware that the behaviour of things or the explanation of events is also complex with many factors at work side by side. He will not be able to see these factors clearly and will not be able, therefore, to find definitive explanations. It would seem that Piaget's thesis has some useful comments to make about teaching method at this time. He suggests that it is not in the nature of developed concrete operations to allow the child to think over a presented situation, to form theories about it, to test them, and to reach conclusions. To do this requires the ability to be aware of the possible factors which may be at work, and to isolate them. Piaget goes on to suggest that this ability is forged, in part, by the child being faced with concrete situations which cannot be readily explained, or by problems which cannot be immediately resolved. That is to say, it is partly in the attempts to resolve concrete problems that formal operations may be assisted in their development. Several points seem to emerge from this from an educational point of view.

Firstly, children need to be faced with situations they cannot readily explain. These situations may arise naturally or may need to be contrived. Secondly, it is thinking over the situation and how to resolve it rather than finding conclusions, which is the important developmental factor. Thirdly, the solutions reached cannot be generalizations, but rather likelihoods or an enumeration of possibilities.

In terms of the classroom, this implies discussions about

possible causes or reasons; finding ways of verifying or refuting suggested reasons; careful observations and recording of data; and open-ended results. Perhaps it could be said that 'experiments in order to see', at this stage, should result in finding many conclusions, which are more a resolution of factors than a precise solution. For example, the question might arise as to why a seedling, which the children had used to study growth rates, had died suddenly. What could be the reasons for this? (Discussion of factors.) A list of relevant and irrelevant factors will emerge. Since we cannot look at all these possibilities, which shall we choose? Careful teacher guidance will be needed here to select one or two factors which the children will be able to isolate experimentally (usually by inversion). How can we make a test to find out? (Methods of resolving the problem.) An experiment could then be worked out which would not be rigorous in the scientific sense, and from which sure conclusions could not be drawn. The conclusion to this study might be, 'we think that John watered the plants too much. We do not think it was the cold weather, because the other plants did not die, but it may also have been lack of plant food in the soil.'

This kind of thinking can be employed in less direct situations, the basis being some fact requiring explanation. What could be the reason? (Discussion.) Enumeration of possibilities, some discarded by discussion. How could we find out? (Research for explanations.) Enumeration of likely reasons. For example, why is our railway station sited where it is? or why is the church made partly of stone and partly of brick?

Early in the period of formal operations, the teaching approach would need to be very like that of the late concrete operational period: the study of data extracted from concrete situations. However, the emphasis could be more upon data relationships, symbolically expressed. Proportion, ratio, probability, percentages, equilibrium of forces, volume, density, etc., could be manipulated abstractly. Experimental work would now have more rigour and the conclusions derived from it would be more definitive. From a teaching point of view, the discussion of hypotheses before action, and the discussion of the results before drawing conclusions, would be a valuable aspect of the work. In terms of the scientific subjects, this would not imply the presentation of a generalization to be verified and then applied in specific cases. But rather the examination of instances (starting perhaps from a discordance), extraction of possible factors, experiment, derivation of a generalization, followed by applications. This might produce sometimes an hypothesis which the teacher knows to be false. This need not be discarded but its falsehood proved along with the truth of other premises. Clearly, children cannot be taught to think 'scientifically' by telling, at this time, any more than they can be made to see that two and two is the same as three and one, during the intuitive period. Nevertheless, teaching methods of this kind do afford an opportunity for the exercise of nascent formal operations. With progress through this period, the children can consider the use of experimental controls and these can be introduced with meaning, i.e. the examination of causal factors by equalization as well as exclusion.

Although a subject like history would still need concrete connotations, the children's activity could extend beyond the description of events. The factors at work within an event might be examined to find the possible reasons for the way it developed. Certain recurrent patterns might be sought out across periods, e.g. battle strategies, chains of authority, and charismatic figures. Role playing might be an effective method in some cases.

Later in the period of formal operations emphasis could be made more upon the possibility than upon the reality. That is to say, generalizations could be applied rather than abstracted. The special case could be considered in this context. The need for contact with the concrete remains in order that potential generalizations may be modified. Direct experience of decision taking and the accompanying interplay of personal attitudes might assist in the study of the humanities. Historical and social studies might be examined from the point of view of event ingredients. Generalizations might be extracted and re-applied to find their value, or otherwise. For example, the patterns of revolution, or conflict, could be looked at across periods, or poverty could be studied in present and past contexts. Events and their consequences might be examined across periods.

This Section has been the briefest outline of some possible relationships between Piaget's work and educational practice. It may have put a little flesh upon the dry bones of the earlier Sections. It remains to be said that the implications of Piagetian psychology are much greater than this account suggests, and can only be discovered by reading his work directly and by verifying possibilities in practice. Piaget's view of the purpose of education, indicated in the following quotation, may breathe life into this skeletal outline and is surely one which no educator would deny:

> The principal goal of education is to create men who are capable of doing new things, not simply of repeating what other generations have done—men who are creative, inventive, and discoverers. The second goal of education is to form minds which can be critical, can verify, and not accept everything they are offered. The great danger today is of slogans, collective opinions, ready-made trends of thought. We have to be able to resist individually, to criticize, to distinguish between what is proven and what is not. So we need pupils who are active, who learn early to find out by themselves, partly by their own spontaneous activity and partly through material we set up for them; who learn early to tell what is verifiable and what is simply the first idea to come to them. (*Piaget Rediscovered,* p. 5.)

Problems and difficulties

It has not been the purpose of this brief outline of Piaget's psychology to make a critical assessment of his work, but rather, a straightforward account of some of his research and theory. In Section 3, his psychology was applied to the educational situation. However, it remains to be said that some of the conclusions which Piaget reaches do not find universal acceptance among psychologists working in this field. This being so, the extrapolations may be misleading. Added to this Section 3 may have placed a wrong interpretation upon Piaget's findings, and would be therefore doubly misleading. In view of this, the following comments will add a note of caution to the extrapolations of this Section.

It has been noted that Piaget regards mental structures as equilibrated systems, the formation of which depends upon the maturation of the nervous system, experience acquired in interaction with the physical environment, and the influence of the social milieu. This being so, the effect which any one of these factors can have upon the structures cannot be considered separately. The point was made on p. 92 et seq., that the equilibrium principle is closely related to the concept of readiness for learning. Readiness for this or that educational experience is another way of saying that the equilibrated structures can accommodate to a given experience. However, if intellectual activity is not best described by means of an equilibrium model, then this view of readiness will not stand. J. Smedslund discusses the arguments for and against equilibration theory in relation to the conservation of substance and weight, and tentatively concludes that the respective validities of learning theory and equilibration theory remain undetermined, although many findings seem to point against the former. (In the *Scandinavian Journal of Psychology*, 1961, 2.) Given that Piaget's variables all play a part, it may be that some factors are much more influential in bringing about structural change than others. In some of his writings, Piaget does suggest that the cultural milieu plays a significant role. Writing about formal operations, in collaboration with B. Inhelder, he comments: 'It follows that their realization can be accelerated or retarded as a function of cultural and educational conditions.' (*Logical Thinking,* p. 337.)

Subsumed under Piaget's heading, 'the social milieu', are several variables, such as language, play experience, general learning, and the formal educational experience. Some pyschologists would place great emphasis upon these variables as structure forming agents. For example, L. S. Vygotsky comments that, '. . . the speech structures mastered by the child become the basic structures of his thinking . . . Thought development is determined by language, i.e. by the linguistic tools of thought and by the sociocultural experience of the child.' (*Thought and Language,* p. 51, the M.I.T. Press, Cambridge, Massachusetts.)

J. S. Bruner takes the view that language plays an important role in conservation acquisition. (*Studies in Cognitive Growth,* page 223, John Wiley & Sons, New York.) Many psychologists have found differences in children's performance with conservation tasks which they feel may be attributable to different cultures and sub-cultures. Other psychologists have found that training programmes can effect structural development. I. E. Sigel, A. Roeper, and F. H. Hooper report a pilot research in which they conclude that training programmes focusing on prerequisites for relevant cognitive operations influence the resultant cognitive structures. (*The British Journal of Educational Psychology,* 1966, 36.) It may be that the autonomy of Piaget's developmental sequence is not inviolable. If this is the case, then to wait for the child to be ready is to mark time, for readiness becomes, to some extent, a variable which is in the hands of the teacher and not the child. Piaget makes the following comment on a question closely related to this:

In some cases, what is transmitted by instruction is well assimilated by the child because it represents in fact an extension of some spontaneous constructions of his own. In such cases, his development is accelerated. But in other cases, the gifts of instruction are presented too soon or too late, or in a manner that precludes assimilation because it does not fit in with the child's spontaneous constructions. Then the child's development is impeded, or even deflected into barrenness, as so often happens in the teaching of the exact sciences. Therefore I do not believe, as Vygotsky seems to do, that new concepts, even at school level, are always acquired through adult didactic intervention. This may occur but there is a much more productive form of instruction: the so-called 'active' schools endeavour to create situations that, while not 'spontaneous' in themselves, evoke spontaneous elaboration on the part of the child, if one manages both to spark his interest and to present the problem in such a way that it corresponds to the structures he had already formed himself. (*Comments,* p. 11.)

There are two other difficulties inherent in Piaget's theory which should be mentioned. It has been noted that the emergence of concrete operations is not a sudden occurrence, and that the operations generalize gradually to various experiential situations. This staggered sequence also shows itself during the period of formal operations. J. F. Wohlwill comments about this matter, in relation to concrete operations, as follows: 'Whether the role of these task variables and the larger problem of the generalizability of a principle or concept can be adequately handled within the framework of Piaget's model of logical operations remains to be seen.' ('Piaget's System as a Source of Empirical Research', in *The Merrill-Palmer Quarterly*, 1963, 4.)

Finally, it should be mentioned that there are problems connected with Piaget's psycho-logic. (C. Parsons, 'Inhelder and Piaget's The Growth of Logical Thinking: II. A logician's viewpoint'. *British Journal of Psychology*, 1960, 51.)

It would be quite wrong to end this short outline of Professor Jean Piaget's massive contribution to child psychology on a discordant note. Can there be any doubt that his unique and remarkable work will always be a predominant milestone in the history of child psychology?

Index